CONTINUOUS-TIME DELTA-SIGMA MODULATORS FOR HIGH-SPEED A/D CONVERSION

Theory, Practice and Fundamental Performance Limits

THE KLUWER INTERNATIONAL SERIES IN ENGINEERING AND COMPUTER SCIENCE

ANALOG CIRCUITS AND SIGNAL PROCESSING

Consulting Editor: **Mohammed Ismail**. *Ohio State University*

Related Titles:

LEARNING ON SILICON: ***Adaptive VLSI Neural Systems***, *Gert Cauwenberghs, Magdy A. Bayoumi;* ISBN: 0-7923-8555-1

ANALOG LAYOUT GENERATION FOR PERFORMANCE AND MANUFACTURABILITY, *Koen Larnpaert, Georges Gielen, Willy Sansen*; ISBN: 0-7923-8479-2

CMOS CURRENT AMPLIFIERS, *Giuseppe Palmisano, Gaetano Palumbo, Salvatore Pennisi*: ISBN: 0-7923-8469-5

HIGHLY LINEAR INTEGRATED WIDEBAND AMPLIFIERS***: Design and Analysis Techniques for Frequencies from Audio to RF***, *Henrik Sjöland*: ISBN: 0-7923-8407-5

DESIGN OF LOW-VOLTAGE LOW-POWER CMOS ΔΣA/D CONVERTERS, *Vincenzo Peluso, Michiel Steyaert, Willy Sansen*: ISBN: 0-7923-8417-2

THE **DESIGN OF LOW-VOLTAGE, LOW-POWER SIGMA-DELTA MODULATORS**, *Shahriar Rabii, Bruce A. Wooley*; ISBN: 0-7923-8361-3

TOP-DOWN DESIGN OF HIGH-PERFORMANCE SIGMA-DELTA MODULATORS, *Fernando Medeiro, Angel Pérez-Verdú, Angel Rodríguez-Vázquez*; ISBN: 0-7923-8352-4

DYNAMIC TRANSLINEAR AND LOG-DOMAIN CIRCUITS: ***Analysis and Synthesis***, *Jan Mulder, Wouter A. Serdijn, Albert C. van der Woerd, Arthur H. M. van Roermund*; ISBN: 0-7923-8355-9

DISTORTION ANALYSIS OF ANALOG INTEGRATED CIRCUITS, *Piet Wambacq, Willy Sansen*; ISBN: 0-7923-8186-6

NEUROMORPHIC SYSTEMS ENGINEERING: ***Neural Networks in Silicon***, *edited by Tor Sverre Lande*; ISBN: 0-7923-8158-0

DESIGN OF MODULATORS FOR OVERSAMPLED CONVERTERS, *Feng Wang, Ramesh Harjani,* ISBN: 0-7923-8063-0

SYMBOLIC ANALYSIS IN ANALOG INTEGRATED CIRCUIT DESIGN, *Henrik Floberg,* ISBN: 0-7923-9969-2

SWITCHED-CURRENT DESIGN AND IMPLEMENTATION OF OVERSAMPLING A/D CONVERTERS, *Nianxiong Tan,* ISBN: 0-7923-9963-3

CMOS WIRELESS TRANSCEIVER DESIGN, *Jan Crols, Michiel Steyaert*, ISBN: 0-7923-9960-9

DESIGN OF LOW-VOLTAGE, LOW-POWER OPERATIONAL AMPLIFIER CELLS, *Ron Hogervorst, Johan H. Huijsing*, ISBN: 0-7923-9781-9

VLSI-COMPATIBLE IMPLEMENTATIONS FOR ARTIFICIAL NEURAL NETWORKS, *Sied Mehdi Fakhraie, Kenneth Carless Smith*, ISBN: 0-7923-9825-4

CHARACTERIZATION METHODS FOR SUBMICRON MOSFETs, edited by *Hisham Haddara*, ISBN: 0-7923-9695-2

LOW-VOLTAGE LOW-POWER ANALOG INTEGRATED CIRCUITS, edited by *Wouter Serdijn*, ISBN: 0-7923-9608-1

INTEGRATED VIDEO-FREQUENCY CONTINUOUS-TIME FILTERS: ***High-Performance Realizations in BiCMOS***, *Scott D. Willingham, Ken Martin*, ISBN: 0-7923-9595-6

FEED-FORWARD NEURAL NETWORKS: ***Vector Decomposition Analysis, Modelling and Analog Implementation,*** *Anne-Johan Annema*, ISBN: 0-7923-9567-0

FREQUENCY COMPENSATION TECHNIQUES LOW-POWER OPERATIONAL AMPLIFIERS, *Ruud Easchauzier, Johan Huijsing*, ISBN: 0-7923-9565-4

ANALOG SIGNAL GENERATION FOR BIST OF MIXED-SIGNAL INTEGRATED CIRCUITS, *Gordon W. Roberts, Albert K. Lu*, ISBN: 0-7923-9564-6

INTEGRATED FIBER-OPTIC RECEIVERS, *Aaron Buchwald, Kenneth W. Martin,* ISBN: 0-7923-9549-2

MODELING WITH AN ANALOG HARDWARE DESCRIPTION LANGUAGE, *H. Alan Mantooth,Mike Fiegenbaum*, ISBN: 0-7923-9516-6

CONTINUOUS-TIME DELTA-SIGMA MODULATORS FOR HIGH-SPEED A/D CONVERSION

Theory, Practice and Fundamental Performance Limits

by

James A. Cherry
Philsar Electronics, Inc

and

W. Martin Snelgrove
Philsar Electronics, Inc.

KLUWER ACADEMIC PUBLISHERS
Boston / Dordrecht / London

Distributors for North, Central and South America:
Kluwer Academic Publishers
101 Philip Drive, Assinippi Park
Norwell, Massachusetts 02061 USA
Telephone (781) 871-6600
Fax (781) 871-6528
E-Mail <kluwer@wkap.com>

Distributors for all other countries:
Kluwer Academic Publishers Group
Distribution Centre
Post Office Box 322
3300 AH Dordrecht, THE NETHERLANDS
Telephone 31 78 6392 392
Fax 31 78 6546 474
E-Mail <orderdept@wkap.nl>

Electronic Services <http://www.wkap.nl>

Library of Congress Cataloging-in-Publication Data

Cherry, James A., 1970-
Continuous-time delta-sigma modulators for high-speed A/D/conversion : theory, practice, and fundamental performance limits / by James A. Cherry and W. Martin Snelgrove.
p. cm. -- (Kluwer international series in engineering and computer science ; SECS 521)
Includes bibliographical references (p.).
ISBN 0-7923-8625-6 (alk. paper)
1. Analog-to-digital converters. 2. Modulators (Electronics) 3. Continuous-time filters. I. Snelgrove, W. Martin, 1954- . II. Title. III. Series
TK7887.6.C49 1999
621.3815'36--dc21 99-40715
CIP

Printed on acid-free paper.

Printed in the United States of America

To Melissa

Contents

List of Figures

List of Tables

List of Abbreviations and Symbols

Abbreviations

A/D	Analog-to-digital
ADC	Analog-to-digital converter; analog-to-digital conversion
AGC	Automatic gain compensation
BJT	Bipolar junction transistor
BP	Band pass
BV_{CEO}	Collector-emitter breakdown voltage with base open
CM	Common-mode
CT	Continuous time
DAC	Digital-to-analog converter
DEM	Dynamic element matching
DFT	Discrete Fourier transform
DNL	Differential nonlinearity
DPW	DAC pulse width
DR	Dynamic range
$\Delta\Sigma$M	Delta–sigma modulator; delta–sigma modulation
DSP	Digital signal processing
DT	Discrete time
ECL	Emitter-coupled logic
FFT	Fast Fourier transform
GB	Gain-bandwidth
HBT	Heterojunction bipolar transistor
HNRZ	Half-delayed non-return-to-zero
HRZ	Half-delayed return-to-zero
IBN	In-band noise
i.i.d.	Independent and identically-distributed
IIP_3	Input third-order intercept point

INL	Integral nonlinearity
LNA	Low-noise amplifier
LP	Low pass
M/S	Master/slave
MSA	Maximum stable amplitude
NRZ	Non-return-to-zero
NTF	Noise transfer function
OOBG	Out-of-band gain
OSR	Oversampling ratio
pdf	Probability distribution function
RHS	Right-hand side
RK4	Fourth-order Runge-Kutta numerical integration
rms	Root mean square
RZ	Return-to-zero
SC	Switched capacitor
SFDR	Spurious-free dynamic range
S/H	Sample and hold
SI	Switched current
SiGe	Silicon germanium
SNDR	Signal-to-noise-plus-distortion ratio
SNR	Signal-to-noise ratio
SQNR	Signal-to-quantization-noise ratio
STF	Signal transfer function
VCO	Voltage-controlled oscillator
ZCT	Zero-crossing time

Roman Symbols

Equation and figure numbers in the list below refer to equations or figures where an example of the use of the symbol may be seen.

a_k	(1.12) Coefficient of $x(n-k)$
A	(Figure 7.5) Imbalance of transistors in multi-tanh block
A_0	(3.1) Op amp dc gain
A_0	Resonator center frequency gain
A_{clk}	(7.11) Clock voltage magnitude
A_i	(5.1) Gain of block $\hat{G}_i(s)$
b	Input tone bin number
b_k	(1.12) Coefficient of $y(n-k)$
B_i	(4.17) Zero-placement feedback coefficients
c_k	(1.12) Coefficient of $u(n-k)$
C	(1.15) Capacitance

C	(6.13) BP resonator tank capacitance
C_μ	Base-collector capacitance
D	Number of emitter diodes in multi-tanh block
e	(Figure 1.2) Quantization noise in linear model
$e(n)$	(5.11) Jitter-induced error sequence
$E(z)$	(1.1) $\mathcal{Z}$-transform of e
f	Frequency
f_c	(5.21) Carrier frequency
f_{max}	Transistor unity power gain frequency
f_n	(5.22) Offset from f_c at which phase noise is measured
f_N	(1.4) Nyquist frequency
f_s	(1.4) Sampling frequency
f_T	(4.2) Transistor unity current gain frequency
g_m	(1.15) Transconductance
G_g, G_q	(6.13) Transconductance of gain and Q-tuning
$\hat{G}(s)$	(1.14) CT prefilter transfer function
$\hat{G}_i(s)$	(5.1) Generalized CT gain block
$G(z)$	(1.11) DT prefilter transfer function blocks
$h(n)$	(2.3) DT loop transfer function impulse response
$\hat{h}(t)$	(2.3) CT loop transfer function impulse response
$\hat{H}(s)$	(1.14) CT loop transfer function
$H(z)$	(1.1) DT loop transfer function
$H(z, \tau_d)$	(4.11) DT loop transfer function of CT $\hat{H}(s)$ with excess loop delay τ_d
I_{Dout}	(Figure 7.21) DAC output current
I_g	(6.13) Gain transconductor output current
I_k	(7.6) Feedback DAC current
I_{le}	(Figure 7.16b) Current through R_{le}
I_q	(6.13) Q-enhancement transconductor output current
I_{tail}	(Figure 7.5) Tail current in multi-tanh block
k	Boltzmann's constant
k_1, k_2	(1.15) Second-order LP $\Delta\Sigma$M feedback parameters
k_2, k_4	(Figure 6.1) BP modulator feedback parameters
k_{h2}, k_{h4}	(Figure 6.1) HRZ feedback coefficients in BP modulator
k_i	(4.17) General feedback coefficients
k_{ni}	(4.29) NRZ feedback coefficients
k_{r2}, k_{r4}	(Figure 6.1) RZ feedback coefficients in BP modulator
$\bar{k}_{r2}, \bar{k}_{r4}, \bar{k}_{h2}, \bar{k}_{h4}$	(6.20) Nominal feedback current levels
k_{ri}, k_{hi}	(4.22) RZ/HRZ feedback coefficients
K	(Figure 1.10) Number of averaged periodograms
l	Loop filter pole multiplicity

L	(6.13) BP resonator tank inductance
L_p	(6.3) Equivalent tank parallel inductance
m	(Figure 1.6) Order of ΔΣM loop filter
n	(1.6) Sampling instant number, $t = nT_s$
n_c	(5.21) VCO phase noise
n_t	(4.2) Number of transistors in feedback path
N	(1.6) Length of output bit sequence
NTF(z)	(1.2) Noise transfer function
p	(3.1) Leaky integration proportionality constant
$p_x(\alpha)$	(Figure 5.14b) pdf of quantizer input voltage
$\hat{P}(f)$	CT-domain power spectrum
$P(n)$	(1.7) DT-domain power spectrum, $n = 0, 1, \ldots, N/2$
P_n	(1.9) Total noise power in baseband
Q	Quality factor
Q_L	(6.2) Inductor quality factor
Q_{res}	Resonator quality factor
Q_n	Bipolar transistor n
R	(6.13) Resistance of inductor in resonant tank
$\hat{r}_D(t)$	(2.3) Time domain feedback DAC impulse response
$\hat{R}_D(s)$	(2.2) Frequency domain feedback DAC impulse response
R_{DAC}	(Figure 7.21) DAC current-source resistance
R_{ef}	(Figure 7.16a) Emitter follower resistance
R_{in}	Input resistance
R_{la}	(Figure 7.16b) Latch amplifier load resistance
R_{le}	(Figure 7.16b) Latch current-source transistor emitter resistance
R_p	(6.3) Equivalent tank parallel resistance
R_{pa}	(Figure 7.16a) Preamp amplifier load resistance
R_{pe}	(Figure 7.16a) Preamp current-source transistor emitter resistance
s	Continuous-time frequency variable, $s = \sigma + j\omega$
s_k	(2.6) kth pole of $\hat{H}(s)$
STF(z)	(1.2) Signal transfer function
t	Time
Δt_{jitt}	(7.12) Time jitter
t_n	(5.4) Time at sample n including jitter
T	Absolute temperature
T_s	Sampling period
u	(1.12) DT input to ΔΣM
$\hat{u}$	(1.15) CT input to ΔΣM

u_{min}, u_{max}	(6.5), (6.6) Minimum, maximum BP modulator input signals
$\hat{U}(s)$	(1.14) Laplace transform of $\hat{u}$
$U(z)$	(1.1) $\mathcal{Z}$-transform of u
$\overline{v_{nc}}$	(7.14) Clock transistor input-referred noise voltage
$\overline{v_{ng1}}$	(6.4) Input-referred noise for G_{g1}
Δv_{noi}	(7.12) Total clock waveform noise
v_{sl}	Slope of latch input voltage relative to full scale
v_x	Latch input voltage relative to full scale
V_{BUF}	(Figure 7.19) Modulator output swing control voltage
V_{CB}	(Figure 7.A.3a) Latch current control voltage
V_{CC}	Power supply voltage
V_{CDB}	(Figure 7.12) V_{IB} control voltage
V_{clk}	(7.11) Clock voltage waveform
V_G	(Figure 7.12) Voltage to generate G_g
V_{h2}, V_{h4}	HNRZ/HRZ DAC control voltages
V_{IB}	(Figure 7.12) Q-tuning multi-tanh block common-mode voltage
V_{k+}, V_{k-}	(Figure 7.21) DAC control voltages
V_{Lmid}	(Figure 7.16b) Intermediate latch voltage
V_{Lout}	(Figure 7.16b) Latch output voltage
V_{n2}, V_{n4}	NRZ DAC control voltages
V_{Pin}	(Figure 7.16a) Preamp input voltage
V_Q	(Figure 7.12) Voltage to generate G_q
V_{r2}, V_{r4}	RZ DAC control voltages
V_{rd}	(Figure 5.10) Differential regeneration voltage
$w(n)$	(1.8) Spectral windowing function
x	(1.12) DT input to quantizer
$\hat{x}$	(2.1) CT input to quantizer
$\hat{x}_1$	(1.15) CT output of first integrator in double integration modulator
$\hat{x}_2$	(1.15) CT output of second integrator in double integration modulator
x_2, x_4	(Figure 6.1) Band pass resonator outputs
X_2	(6.13) Laplace transform of x_2
x_i	(5.3) Output of gain block $\hat{G}_i(s)$
$\hat{X}(s)$	(1.14) Laplace transform of $\hat{x}$
$X(z)$	(1.11) $\mathcal{Z}$-transform of x
y	(1.12) DT output of $\Delta\Sigma$M
$\hat{y}$	(Figure 2.1a) CT output of $\Delta\Sigma$M
$Y(n)$	(1.6) DFT of y

$\hat{Y}(s)$	(1.14) Laplace transform of $\hat{y}$
$Y(z)$	(1.1) $\mathcal{Z}$-transform of y
z	(1.1) Discrete-time frequency variable, $z = \exp(sT)$
z_k	(2.6) kth pole of $H(z)$

Greek Symbols

(α, β)	(2.4) DAC pulse interval
β	Random variable for jitter distribution
β_n	(5.4) Jitter at sample n
δy	Difference between successive output samples, $y(n) - y(n-1)$
Δ	Difference between adjacent quantizer output levels
$\epsilon_{g1}, \epsilon_{g2}$	(6.7) Nonlinearity coefficients of G_{g1}, G_{g2}
$\bar{\epsilon}_{g2}$	(6.28) Normalized ϵ_{g2}
$\epsilon_{q1}, \epsilon_{q2}$	Nonlinearity coefficients of G_{q1}, G_{q2}
$\bar{\epsilon}_{q1}, \bar{\epsilon}_{q3}$	(6.27), (6.29) Normalized $\epsilon_{q1}, \epsilon_{q3}$
$\|\|\varepsilon\|\|_2$	(2.43) RMS least-squares matching error
$\|\|\varepsilon\|\|_\infty$	(2.43) Maximum least-squares matching error
γ_k	(6.23) DAC feedback current scale factor
κ	(Figure 4.4) Linearized quantizer gain
ρ_d	(4.1) Excess loop delay as a fraction of T_s, i.e., τ_d/T_s
ρ_r	DAC pulse rise time as a fraction of T_s, i.e., τ_r/T_s
σ	Standard deviation
σ_β	(5.12) Standard deviation of β
$\sigma_{\delta y}$	(5.12) Standard deviation of δy
σ_e	(5.12) Standard deviation of $e(n)$
σ_x	Standard deviation of $x(n)$
$\sigma_{x_2}, \sigma_{x_4}$	(6.25) Standard deviation of x_2, x_4
τ	Op amp dominant pole time constant
τ_d	(4.1) Excess loop delay
τ_r	DAC pulse rise time
τ_{rg}	Latch regeneration time constant
ω_0	Center frequency of band pass resonator
ω_i	(5.2) Center frequency of gain block $\hat{G}_i(s)$

Preface

The delta–sigma modulator, a single-bit oversampled analog signal encoder, was patented by Cutler in 1960 and described in the open literature by Inose and Yasuda in 1962. It took until the mid-1980s, specifically the publication of Candy's widely-cited 1985 paper about the double integration modulator, for the popularity of these modulators to take off. Since then, almost fifteen years later, delta–sigma modulators have cornered the market in digital audio converters (both analog-to-digital and digital-to-analog), and they are also used to great advantage in lower-rate but still higher-resolution applications such as instrumentation and seismometers.

Most modulators are implemented with a discrete-time loop filter using circuit techniques such as switched-capacitor, likely because it is easy to map the mathematics of modulators onto the implementation. Due to settling-time constraints in typical discrete-time implementations, the maximum clock rate is limited, and because delta–sigma circuits are necessarily oversampled, high-resolution converters with only relatively narrow conversion bandwidths (up to a few hundred kilohertz) can be built. Ideally, resolution is determined only by the modulator loop filter and the oversampling ratio; if it were possible to relax the clock rate restrictions and thereby increase the clock rate, then for a given oversampling ratio and loop filter, high-resolution converters with wider bandwidths (into the megahertz) could be built.

One way to achieve higher clock rates is to build the loop filter portion out of continuous-time circuits such as transconductors and integrators. The idea of using continuous-time filters in delta–sigma modulators is not particularly new: Adams' seminal 1986 paper describes one of the first functional delta–sigma audio encoders that had a continuous-time loop filter built out of discrete op amps, resistors, and capacitors. However, since the mid-1990s, semiconductor processes have increased in speed to the point that it becomes possible to integrate gigahertz-speed modulators. An example of an application where such modulators are especially appealing is cellular radio: an entire cellular band (tens of megahertz) could be digitized, possibly at the IF or RF, and the signal details sorted out with digital signal processing.

Norsworthy, Schreier, and Temes' excellent 1997 IEEE Press book "Delta–Sigma Data Converters: Theory, Design, and Simulation" deals with a host of important subjects in discrete-time modulator design. We have attempted to make this monograph complementary to their book: we treat continuous-time modulator design in detail, emphasizing high-speed modulators in particular. Our monograph is intended for engineers generally interested in fast oversampled converters: we cover theoretical aspects of design (such as discrete-to-continuous transformation and how to measure modulator performance in a simulation), and we also provide a wealth of practical information, such as common continuous-time modulator nonidealities, a survey of published high-speed modulators, and detailed test results on a manufactured high-speed part. Although we hope our book provides enough detail to be understood on its own, we do not hesitate to recommend "Delta–Sigma Data Converters" as a source of additional information; we will make frequent reference to it.

Chapter 1 of this monograph introduces preliminary material important in modulator design and understanding. We provide a brief introduction to the key concepts in delta–sigma modulation including noise-shaping, oversampling, and fundamental modulator design choices. We briefly state some advantages continuous-time modulators have over their discrete-time counterparts. As well, we list important modulator performance measures and give explicit instructions on how to find them, something we find done infrequently or incompletely elsewhere. Finally, we highlight some issues surrounding the time-domain simulation of delta–sigma modulators.

Chapter 2 covers the design of continuous-time delta–sigma modulators. We first treat ideal modulator design: the impulse-invariant transformation for mapping discrete filter transfer functions to equivalent continuous filters is explained, the multi-feedback band pass modulator architecture is shown, and worked examples are provided. The built-in antialiasing property of continuous-time modulators is illustrated, and the effect of unequal feedback pulse rise and fall times is investigated. We then explain some techniques which may be employed to design and verify the filter characteristics of actual behavioral-level or transistor-level modulators in SPICE or any other computer simulation program.

Chapter 3 explores practical issues in continuous-time modulators. We survey the literature on nonidealities in discrete-time modulators and explain if, and how, these apply in continuous-time modulators. We also list and give brief synopses of important delta–sigma papers in the published literature, and we summarize the performance achieved in high-speed continuous-time modulators.

Chapter 4, along with the following three chapters, describes a specific aspect of high-speed modulator design: delay in the modulator feedback path, denoted excess loop delay. We explain why this delay is problematic, unify and extend past work in the area, and give concrete suggestions for surmounting the effects of excess delay.

Chapter 5 investigates the important topic of clock jitter. We show that it can be more severe in continuous-time modulators than in dicrete-time modulators, and we show the effects both of white clock jitter and of jitter with a phase-noise characteristic typical of an integrated voltage-controlled oscillator—this latter subject is important for any modulator realized in a fully-integrated form. Moreover, we delve into the effects of quantizer metastability on modulator performance: we derive and validate a behavioral model for a metastable quantizer, and use it to study the seriousness of metastability and propose methods for reducing its effects.

Chapter 6 contains an explicit design procedure for a high-speed fourth-order continuous-time band pass modulator, where parameter selection is explained and tradeoffs are identified. Chapter 7 characterizes a fabricated silicon-germanium modulator for conversion of UHF analog signals to digital. We present detailed circuit simulation and test results, and then discuss issues that limit the measured performance and give techniques to overcome them.

Finally, Chapter 8 summarizes the work presented. We consider the challenges in, and appropriateness of, high-speed analog-to-digital conversion using continuous-time delta–sigma modulators, and we list some applications in which we feel delta–sigma modulation is put to good use.

This book is in large part taken from the first author's Ph.D. thesis. We have done much reorganization and rewriting to make the tone more informal and appealing; we hope the material itself remains accessible and instructive.

JAMES A. CHERRY
W. MARTIN SNELGROVE

JUNE 1999

Acknowledgments

Many people contributed directly or indirectly to this work, and we express our gratitude to them.

For research inspiration, thanks to Dr. Omid Shoaei and Dr. Lars Risbo for their thought-provoking Ph.D. theses, Mr. Joseph Jensen and Dr. Gopal Raghavan at HRL for their seminal paper on high-speed $\Delta\Sigma$M, and special thanks to Mr. Weinan Gao for discussions on theory and for allowing us access to his circuit design for simulation and testing purposes. His generosity has given this work a much more practical slant than it otherwise would have had.

At Nortel Networks, thanks to Mr. Robert Hadaway for allowing us access to the SiGe process, Dr. Sorin Voinigescu and Mr. Dave Marchesan for assistance with layout and simulation, M. Étienne Bernard and Ms. Marie Quinn for wafer-dicing expertise, Dr. José Macedo and Mr. Glen Tiegs for helpful packaging suggestions, Ms. Icy Jackson for rapid packaging of test chips, and special thanks to Dr. Peter Schvan for all his help with computer system problems and for offering us the use of his SPICE files for a prototype high-speed modulator.

For modulator testing, thanks to Dr. Sergiy Uhanov and Mr. Nagui Mikhail at Carleton University for help with procuring and calibrating measurement equipment, special thanks to Mr. Joseph Jensen and Dr. Gopal Raghavan (once again!) at HRL for letting us capture bit streams on their high-speed deep-sampling oscilloscope, and extra-special thanks to Mr. Philip J. Lauzon for his tireless design, redesign, and assembling of printed circuit boards.

Finally, thanks to the Natural Sciences and Engineering Research Council of Canada for providing a substantial portion of the funding that allowed this work to be completed.

Chapter 1

DELTA–SIGMA MODULATION CONCEPTS

In this chapter we explain what a delta–sigma modulator ($\Delta\Sigma$M) is and how it can be used for analog-to-digital conversion along with some of the basic design choices in $\Delta\Sigma$M design. Next, we list some of the advantages of continuous-time modulators over discrete-time modulators. We move on to describing how to measure the performance of a $\Delta\Sigma$M, and we close by discussing some aspects of the time-domain simulation of $\Delta\Sigma$Ms, distinguishing discrete-time modulator simulation from continuous-time.

1. A BRIEF INTRODUCTION TO $\Delta\Sigma$M

An overview of the $\Delta\Sigma$M concepts relevant for this work will be presented here. There are a number of other excellent summary articles to which the interested reader may turn for an alternate or more detailed treatment [Hau91, Can92b, Azi96, Can97].

1.1 OPERATING PRINCIPLES

A $\Delta\Sigma$M analog-to-digital converter (ADC) has three important components, depicted in Figure 1.1:

1. A *loop filter* or *loop transfer function* $H(z)$

2. A clocked quantizer

3. A feedback digital-to-analog converter (DAC)

The quantizer is a strongly-nonlinear circuit in an otherwise linear system, which makes the behavior of $\Delta\Sigma$Ms very complicated to investigate analytically [Gra90]. However, the basic idea of $\Delta\Sigma$ modulation is easy enough to state: the analog input signal is modulated into a digital word sequence whose spectrum

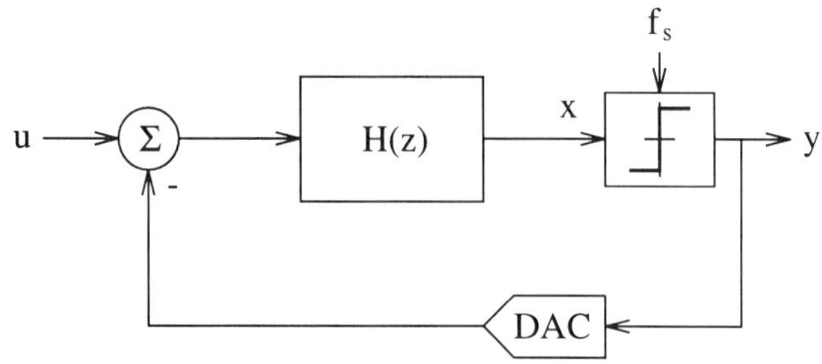

Figure 1.1. Basic components of a ΔΣM for ADC.

approximates that of the analog input well in a narrow frequency range, but which is otherwise noisy. This noise arises from the quantization of the analog signal, and the loop filter "shapes" the quantization noise away from the narrow (desired) frequency range.

An intuitive qualitative understanding of how this happens can be had by *linearizing* the circuit as shown in Figure 1.2. The quantizer is replaced by an

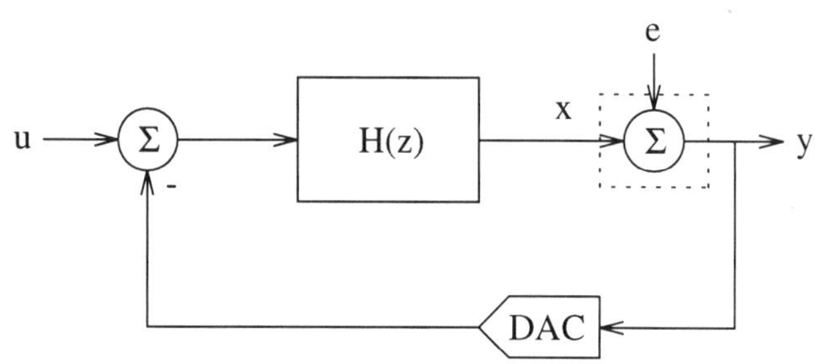

Figure 1.2. Linearizing the quantizer in a ΔΣM.

adder and we pretend that the quantization noise is "generated" by an input e which is independent of the circuit input u. It is as though the quantizer knows exactly the right value to add to its input that gives an output value which falls exactly at one of the discrete output levels. The output y may now be written in terms of the two inputs u and e as

$$Y(z) = \frac{H(z)}{1+H(z)}U(z) + \frac{1}{1+H(z)}E(z) \quad (1.1)$$

$$= \text{STF}(z) \cdot U(z) + \text{NTF}(z) \cdot E(z) \quad (1.2)$$

where STF(z) and NTF(z) are the so-called *signal transfer function* and *noise transfer function*. From (1.1) we see that the poles of $H(z)$ become the zeros of NTF(z), and that for any frequency where $H(z) \gg 1$,

$$Y(z) \approx U(z).$$

In other words, the input and output spectra are in greatest agreement at frequencies where the gain of $H(z)$ is large.

Example 1.1: Consider the system of Figure 1.1 with a simple integrator $H(z) = 1/(z-1)$ as the loop filter and a one-bit quantizer which produces output bits with values ± 1. From (1.1) we can calculate

$$\text{STF}(z) = z^{-1}, \ \text{NTF}(z) = 1 - z^{-1}. \tag{1.3}$$

These are depicted graphically in Figure 1.3 with $z = \exp(j2\pi f T_s)$. We have $H(z) \to \infty$ at dc (i.e., at $f = 0$), which means input signals near dc should be reproduced faithfully in the output bit stream. In fact, $|\text{STF}(z)| = 1$ everywhere, so we at least expect the *magnitude* (if not the phase) of an input at any frequency to be reproduced at the output. As well, $\text{NTF}(z) \to 0$ at dc, and it increases away from dc; hence, we say the quantization noise is "shaped away from dc". .

If we implement the system mathematically, simulate it in Matlab [Han98], and look at the power spectrum of the output bit stream, we obtain the plot shown in Figure 1.4. In this simulation, the input tone had an amplitude of 0.434V and a frequency of $(8.545 \times 10^{-3})f_s$. Relative to the output levels of ± 1, we expect, and observe, an output signal power of $20\log_{10}(0.434/\sqrt{2}) = -10.2$dB. The quantization noise spectrum follows NTF(z) qualitatively at least, going to zero at dc and increasing away from dc, but it clearly contains tones spaced at an interval related to the input frequency. □

The usual assumption when linearizing the quantizer as in (1.1) is that the quantization noise spectrum is white, as well as uncorrelated with the input; while the former is often true, the latter is never exactly true, though the correlation is often so complex as to be all but impossible to express analytically. The linearization is thus not really valid, but it often gives correct qualitative predictions of modulator performance. However, we usually require *quantitative* accuracy, and thus for the most part we eschew quantizer linearization throughout this monograph.

Note what is implied in the previous example: the quantization noise is reduced only in a small bandwidth, that is, a bandwidth much smaller than the sampling frequency f_s. If we wish to obtain high converter resolution, then the signal must be bandlimited to a value much smaller than f_s. This means that for a signal with Nyquist rate f_N, we require $f_N \ll f_s$, which is the same as

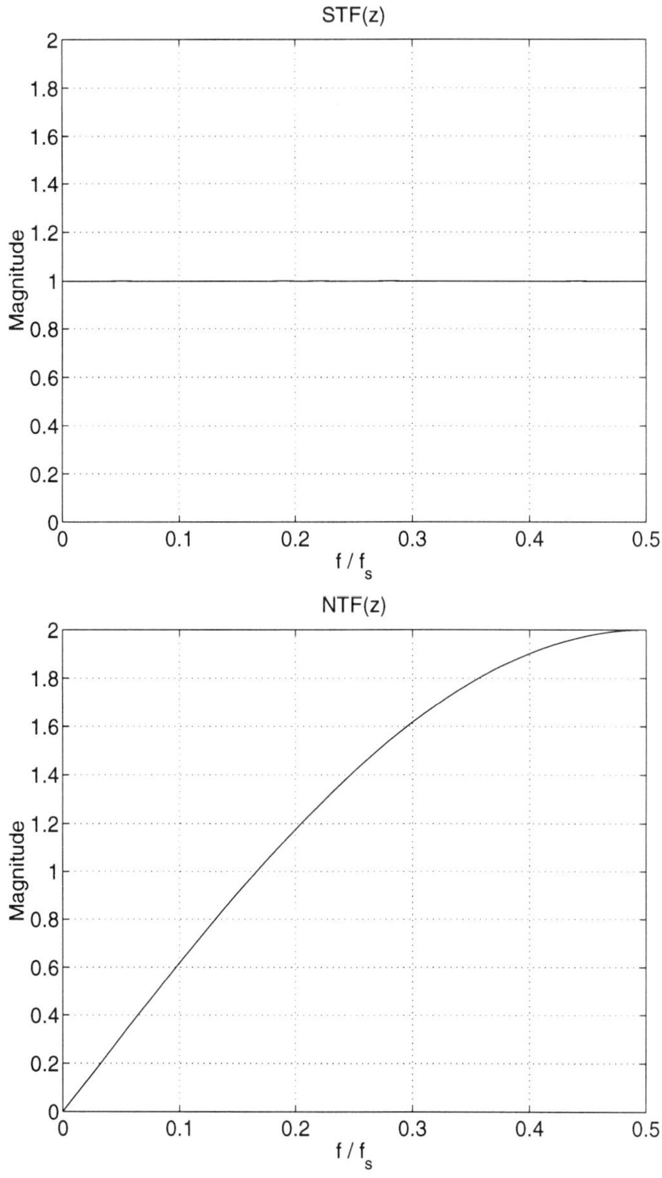

Figure 1.3. STF(z) and NTF(z) for circuit of Example 1.1.

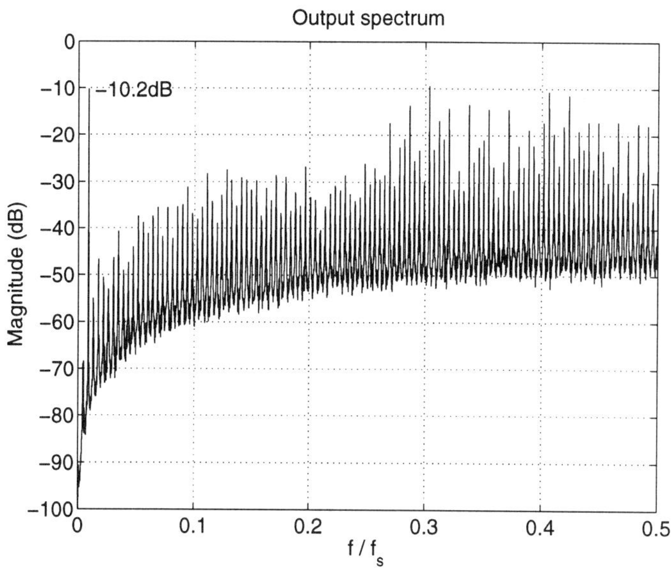

Figure 1.4. Simulated output bit stream power spectrum.

saying we must sample much *faster* than the Nyquist rate. $\Delta\Sigma$Ms, therefore, are so-called *oversampled converters*, with an *oversampling ratio* defined as

$$\text{OSR} \equiv f_s/f_N. \tag{1.4}$$

How is the high-speed low-resolution quantizer output converted to multibit output samples at the Nyquist rate? A complete block diagram of a $\Delta\Sigma$M ADC is shown in Figure 1.5; it includes a modulator followed by a circuit called a *decimator*. The decimator's purpose is twofold: it *decimates* the high-rate bit stream, i.e., reduces it in frequency, *and* removes everything outside the desired band with a filter. Typical time domain and frequency domain waveforms at the modulator and decimator outputs are shown in Figure 1.5.

We shall not go into detail regarding the design of the decimator, instead preferring to concentrate on designing a $\Delta\Sigma$M to obtain an output bit stream with desirable properties. Decimator design is reasonably well-understood and is covered in [Can92a]. As is customarily done in work about $\Delta\Sigma$M, we shall assume that the modulator output is filtered by a brick-wall filter with a gain of 1 in the signal band and 0 elsewhere.

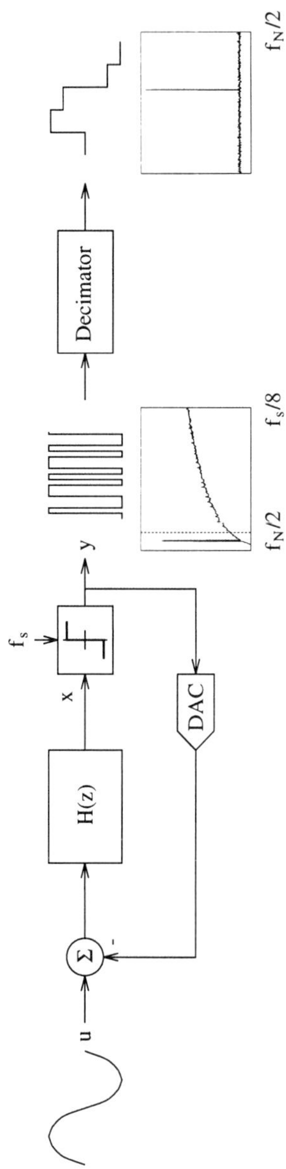

Figure 1.5. Complete $\Delta\Sigma$M ADC block diagram including decimator.

1.2 DESIGN CHOICES

There is a myriad of design choices for ΔΣMs. The major ones are listed and described very briefly here.

MODULATOR ORDER AND OVERSAMPLING RATIO

Example 1.1 featured a single integrator, a first-order transfer function, for $H(z)$. In general, the order of $H(z)$ (which must be strictly proper to ensure causality) is the maximum power of z in the denominator. It is possible to use a second-, third-, or even higher-order $H(z)$ as a loop filter; generally, a converter of order m is built as a cascade of m integrators usually surrounded with feedforward and feedback coefficients. A typical architecture is depicted in Figure 1.6 [Cha90].

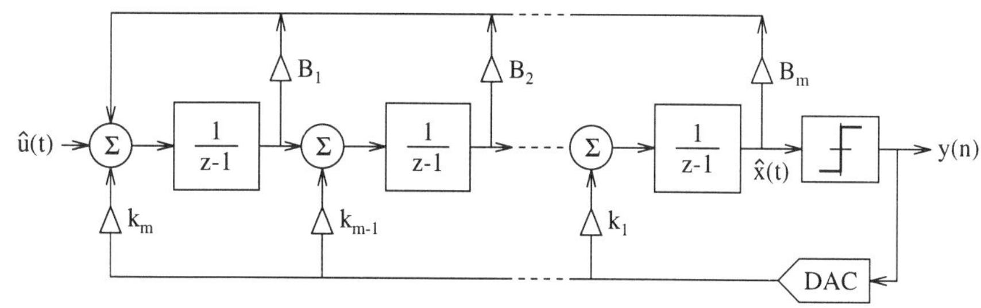

Figure 1.6. General mth-order low pass ΔΣM structure.

In a given application, the signal bandwidth f_N is usually fixed. Sampling faster than the Nyquist rate (i.e., oversampling) is *always* beneficial for improving the measured signal-to-noise ratio (SNR) in an ADC. This is true if the quantization noise inside the signal band is white, as it is in a traditional ADC: doubling the OSR (i.e., increasing it by an octave) halves the bandwidth, and hence the noise power, so that SNR improves by 3dB. The SNR in an order-m ΔΣM improves by $6m + 3$dB per octave of oversampling [Can92b] because the noise is *shaped* by the loop filter. Thus, a high-order modulator is desirable because of the huge increase in converter dynamic range (DR) obtained from a doubling of the OSR.

Not surprisingly, using a high-order modulator has drawbacks. First, the stability of the overall system with $H(z)$ above order two becomes conditional: input signals whose amplitudes are below but close to full scale (to be defined later) can cause overload at the output of the integrators closer to the quantizer, which degrades DR [Sch93]. As well, the placement of the poles and zeros of $H(z)$ becomes a complicated problem, though many solutions have been proposed in the literature (e.g., [Ris94], [Nor97, Chap. 4], among others). Furthermore, the technology in which the circuit is implemented and the circuit architecture itself will limit the maximum-achievable sampling rate and hence, from (1.4), limit the OSR. Finally, the design of the decimator increases in complexity and area for larger oversampling ratios. Typical values of OSR lie in the range 32–256, though circuits with OSRs outside this range have been fabricated [Bai96, Nys96].

QUANTIZER RESOLUTION

It is possible to replace the single-bit quantizer of Example 1.1 with a multi-bit quantizer, e.g., a flash converter [Ada86]. This has two major benefits: it improves overall $\Delta\Sigma$M resolution, and it tends to make higher-order modulators more stable. Furthermore, nonidealities in the quantizer (e.g., slightly misplaced levels or hysteresis) don't degrade performance much because the quantizer is preceded by several high-gain integrators, hence the input-referred error is small [Hau86]. Its two major drawbacks are the increase in complexity of a multibit vs. a one-bit quantizer, and that the feedback DAC nonidealities are directly input-referred so that a slight error in one DAC level reduces converter performance substantially. There exist methods known as *dynamic element matching techniques* to compensate for multibit DAC level errors (e.g., [Gal96], [Lar88]). These aren't needed in a single-bit design because one-bit quantizers are inherently linear [Sch93].

LOW PASS VS. BAND PASS

Integrators have poles at dc, and hence building $H(z)$ from integrators will shape noise away from dc. $\Delta\Sigma$Ms where the quantization noise has a high pass shape are built with low pass loop filters and hence are denoted *low pass* (LP) converters. If we were to build $H(z)$ out of *resonators*, the noise would tend to be shaped away from the resonant frequency. The quantization noise then has a band stop shape because the loop filter is band pass, and the resulting $\Delta\Sigma$Ms are called *band pass* (BP) converters [Sch91]. A common type of band pass converter is built starting with a low pass $H(z)$ and performing the substitution $z^{-1} \rightarrow -z^{-2}$; this produces a converter with noise shaped away from $f_s/4$ with identical stability properties and performance as the low pass prototype, though the order is doubled [Sho96].

A typical application of such a converter is the conversion of an RF or IF signal to digital for processing and heterodyning in the digital domain, as depicted in Figure 1.7. The spectrum at the output of the converter is shown in the figure—the quantization noise is large everywhere except in a narrow band near 1GHz. Mixing to baseband digitally for I and Q channel recovery becomes particularly easy when the sampling frequency is chosen to be four times the input signal frequency because sine and cosine are sequences involving only ± 1 and 0, so simple digital logic can replace a complicated multiplier circuit. In general, the ability of a $\Delta\Sigma$M to perform narrowband

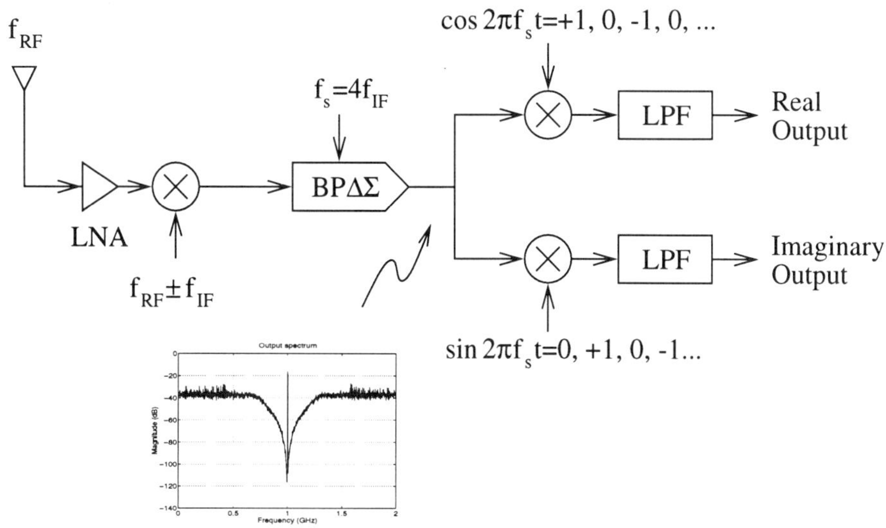

Figure 1.7. Typical radio receiver application for a band pass $\Delta\Sigma$M.

conversion at a frequency other than dc makes them particularly attractive for radio applications; furthermore, as we note below, CT $\Delta\Sigma$Ms can be made fast enough to allow conversion of signals into the hundreds of MHz and beyond[1].

OSR for BP converters is defined as half the sampling frequency divided by the bandwidth of interest [Nor97, Chap. 9]. Thus, an $f_s/4$ converter with a signal occupying the frequency range $(f_s/4 - f_s/32, f_s/4 + f_s/32)$ has a bandwidth of $f_s/16$, and hence OSR $= 8$.

SINGLE STAGE VS. MULTI-STAGE

Many modulators employ a single quantizer with multiple feedback loops leading to various points inside the forward modulator path, and these are called *single-stage* or *multiloop* $\Delta\Sigma$Ms. It is possible to build stable high-order modulators out of two or more low-order modulators where later modulators' inputs are the *quantization noise* from previous stages. Such $\Delta\Sigma$Ms are called *multistage*; they were originally dubbed "MASH" structures, where MASH is an acronym deriving somehow from Multistage Noise-Shaping [Hay86]. In Figure 1.8, a first-order modulator's quantization noise is shaped by another

[1]This is not the only possible architecture: we might digitize directly at the RF rather than at the IF, although the noise figure of the $\Delta\Sigma$M might be too high to achieve the desired system dynamic range. We might also mix more than once prior to the modulator.

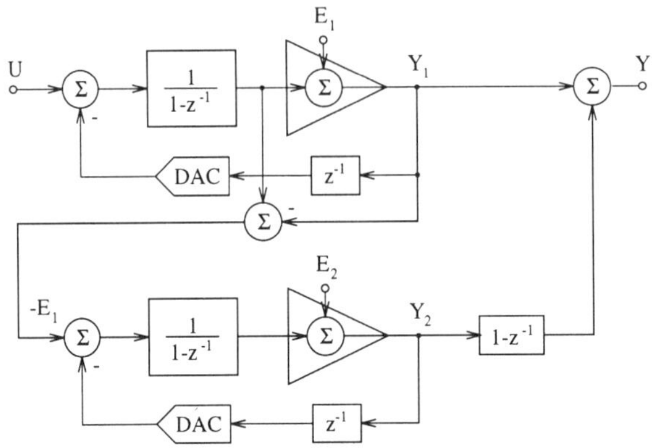

Figure 1.8. A multistage ΔΣM.

first-order modulator:

$$
\begin{aligned}
Y_1 &= U + (1 - z^{-1})E_1 \\
Y_2 &= -E_1 + (1 - z^{-1})E_2.
\end{aligned}
$$

When Y_2 is differentiated and added to Y_1, we find

$$
\begin{aligned}
Y &= Y_1 + (1 - z^{-1})Y_2 \\
&= U + (1 - z^{-1})E_1 - (1 - z^{-1})E_1 + (1 - z^{-1})^2 E_2 \\
&= U + (1 - z^{-1})^2 E_2. \qquad (1.5)
\end{aligned}
$$

Thus, the first-order noise is canceled in the output and the modulator achieves second-order quantization noise shaping. In principle, this can be extended to mth order noise shaping while preserving unconditional stability since each first-order ΔΣM is unconditionally stable. In practice, mismatches between components in the stages result in imperfect noise cancellation [Mat87].

DISCRETE- VS. CONTINUOUS-TIME

We have been writing the loop transfer function $H(z)$ in the discrete-time (DT) domain. The majority of ΔΣMs in the literature are implemented as discrete-time circuits such as switched-capacitor (SC) [Bai96] or switched-current (SI) [Ned95] circuits. It is possible to build the loop filter as a *continuous-time* (CT) circuit $\hat{H}(s)$, for example with transconductors and integrators [Jen95]. There are several reasons why we might choose to do this, as we are about to see.

2. WHY CONTINUOUS-TIME $\Delta\Sigma$M?

Why might we choose to employ a CT filter inside our modulator loop rather than a DT filter? Among the possible reasons are the following.

1. A typical SC DT $\Delta\Sigma$M has a maximum clock rate limited both by op amp bandwidths and by the fact that circuit waveforms need several time constants (i.e., clock periods) to settle. For a $\Delta\Sigma$M built in a process with maximum transistor speed f_T, the maximum clock rate of an SC modulator is on the order of $f_T/100$. By contrast, waveforms vary continuously in a CT $\Delta\Sigma$M, and the restriction on op amp bandwidths (if any op amps are even used) are relaxed. In theory, a CT modulator could be clocked up to an order of magnitude faster in the same technology without much performance penalty. A significant portion of the remainder of this book is devoted to quantifying this statement more precisely.

2. In a DT modulator, large glitches appear on op amp virtual ground nodes due to switching transients. This is not the case in a CT modulator: op amp virtual grounds can be kept very quiet. One example of where this is useful is [dS90], where in order to achieve 20-bit resolution the first of four modulator stages was made out of a CT integrator: the input voltage was applied to a resistor connected to the (clean) virtual ground node, which allowed a current of sufficient accuracy to be generated.

3. One problem with working in the DT domain is *aliasing*: signals separated by a multiple of the sampling frequency are indistinguishable [Gre86, Chap. 2]. DT $\Delta\Sigma$Ms usually require a separate filter at their inputs to attenuate aliases sufficiently. We shall see in Chapter 2 §2. that CT $\Delta\Sigma$Ms have free antialiasing: it can be shown that antialiasing is an inherent property of the mathematics of CT $\Delta\Sigma$Ms.

4. Perhaps the major reason for the prevalence of DT-based $\Delta\Sigma$Ms is that there is a natural allegory between the mathematics of the system and its circuit-level implementation. Fundamentally, $\Delta\Sigma$Ms rely on having at least one integrator inside the loop; in the DT domain, we must usually construct a circuit to give us the $1/(z-1)$ integrating function. It happens that there exist certain devices (like accelerometers [Lu95] and fluxgate magnetic sensors [Kaw96]) which behave as physical continuous-time integrators—they implement a $1/s$ function. Circuits have been built which use these devices as the first stage of a $\Delta\Sigma$M. Thinking in the CT domain, therefore, allows us to realize compact converters outside the voltage/current domain.

We illustrate in Chapter 2 §1.1 that choosing the loop filter for a CT $\Delta\Sigma$M is no more difficult than choosing a DT $\Delta\Sigma$M's loop filter. Thus, CT modulators can be used essentially anywhere a DT modulator can be.

We feel it prudent to note here that it is possible to build MASH modulators with CT loop filters [Nor97, Chap. 6], but to our knowledge, all published MASH $\Delta\Sigma$Ms to date have been DT. As such, we will consider only single-stage modulators in this work.

3. PERFORMANCE MEASURES

We have mentioned certain A/D converter performance measures such as dynamic range and signal-to-noise ratio, but we have yet to explain how to determine them for a $\Delta\Sigma$M. This topic is not often covered in detail in the literature, something we hope to remedy here.

3.1 POWER SPECTRUM ESTIMATION

A $\Delta\Sigma$M is a *noise-shaping converter*: the quantization noise is shaped away from the desired frequency band. We are thus interested in the frequency domain representation of the time domain output bits. More specifically, we care about the power spectrum of the output bits. The most common tool for finding power spectra is the discrete Fourier transform or DFT.

Suppose we have N uniformly-sampled data points $y(n) = \hat{y}(t)|_{t=nT_s}$, $n = 0 \ldots N-1$, $y(n) \in \mathbf{R}$. We will be using the so-called *periodogram* to estimate the power spectrum of $y(n)$. The DFT (which can be implemented using a fast algorithm called the *fast Fourier transform* or FFT when N is a power of two) of $y(n)$ is given by

$$Y(n) = \sum_{k=0}^{N-1} y(k) e^{j2\pi kn/N}, \; n = 0 \ldots N-1 \tag{1.6}$$

and the periodogram is defined as [Pre92]

$$\begin{aligned} P(0) &= \frac{1}{N^2}|Y(0)|^2 \\ P(n) &= \frac{1}{N^2}[|Y(n)|^2 + |Y(N-n)|^2], \; n = 1 \ldots (\tfrac{N}{2}-1) \\ P(N/2) &= \frac{1}{N^2}|Y(N/2)|^2. \end{aligned} \tag{1.7}$$

This power spectrum is defined at $N/2+1$ uniformly-spaced frequency points between 0 and the Nyquist rate $f_s/2$. Thus, each frequency *bin* is of width f_s/N.

An example plot of $10\log_{10} P$ from (1.7) was shown in Figure 1.4 in Example 1.1. Evidently, P is rms power: our input had magnitude -7.2dB and its power in the spectrum is -10.2dB. In this monograph, when we refer to the "spectrum", we mean the power spectrum as found from the periodogram. In practical terms, to find the spectrum of a modulator's simulated or measured output, we recommend Matlab: take N (a power of two, typically in the range 4096 to 32768) bits represented as ± 1 and store them in a vector, then use Matlab's `spectrum` function on that vector of bits.

A periodogram is a discrete representation of the spectrum of a discrete (sampled) signal, but in the real world power spectra are continuous functions of continuous signals. The discretization gives rise to two problems in periodograms, the first of which is usually denoted *spectral leakage* or simply *leakage*, and the second of which relates to uncertainty. We discuss both and how to alleviate them below.

LEAKAGE AND WINDOWING

If there exists a tone in the input signal at a frequency that does not fall exactly in the center of a frequency bin, then leakage will result: instead of a sharp "spike" in one spectrum bin, the tone will become spread over several adjacent bins. This can be understood by realizing that we can only take the FFT of a finite stretch of data (i.e., at a finite number of points); this is akin to taking the FFT of an infinite stretch of data multiplied by a rectangular window that is 1 for the duration of the finite stretch and 0 elsewhere. In the frequency domain, this corresponds to convolving an infinite power spectrum with the Fourier transform of a rectangle, namely, $(\sin x)/x$. The amount of leakage is determined by the spectrum of this function.

The severity of leakage may be reduced by *windowing* the data, which means multiplying it by a *windowing function* before taking its FFT. This has the effect of convolving the spectrum with a function other than $(\sin x)/x$. [Har78] lists many examples of windows; in the time domain, they generally peak at 1 near the center of the data and fall to 0 in various ways near the edges. We prefer to use a *Hann window* (often incorrectly called a *Hanning window*), also called a *raised cosine* window because of the formula that describes it:

$$w(n) = \frac{1}{2}\left[1 - \cos\left(\frac{2\pi n}{N}\right)\right], n = 0, \ldots, N-1. \qquad (1.8)$$

Example 1.2: In the simulation of $\Delta\Sigma$Ms, it is easy (and recommended) to choose an input sinusoid with a frequency exactly in the center of a bin by making its frequency a multiple of f_s/N. Thus, leakage from the input tone is not usually problematic. Moreover, discrete tones arising from output limit cycles also usually fall exactly in the center of frequency bins. However, one case where they don't occurs when simulating a low pass $\Delta\Sigma$M and the mean of $y(n)$ is nonzero. This creates a dc component in $P(n)$ and also "misaligns" the output limit cycles such that there is leakage into all the low-frequency bins. We shall see that this turns out to give an unfairly-pessimistic SNR estimate.

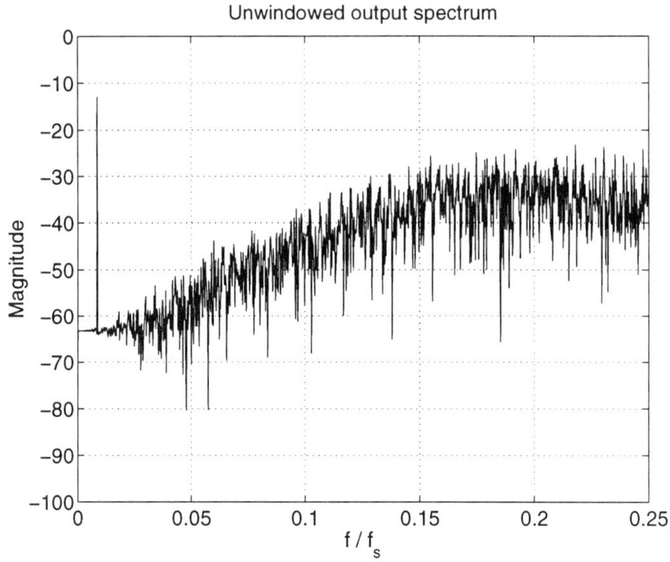

Figure 1.9a. Effect of windowing on unwindowed output spectrum.

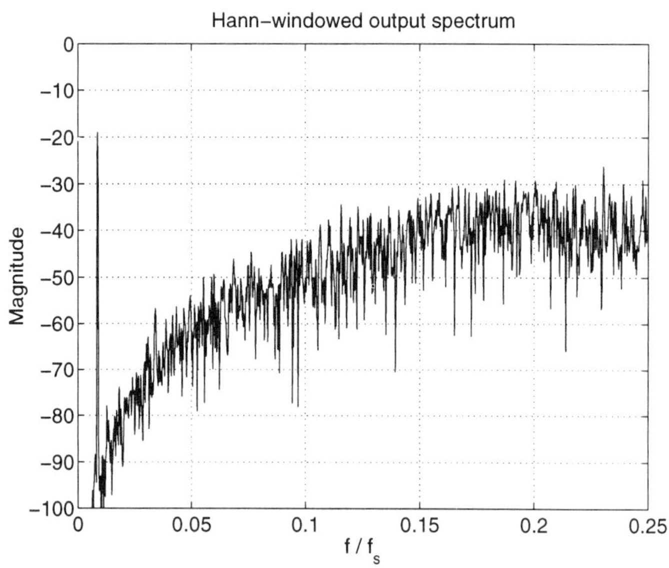

Figure 1.9b. Effect of windowing on windowed output spectrum.

Windowing greatly alleviates the problem. Figure 1.9a illustrates what happens when $N = 4096$ output bits from a second-order modulator have an average value of $2/N = -66.2$dB: the spectrum near dc flattens out to -63.2dB. Taking that same output bit stream and first multiplying it by a Hann window before taking the FFT yields Figure 1.9b: now, the noise-shaping behavior is clearly evident down to dc. We prefer a Hann window because the input tone only becomes smeared over its immediately adjacent bin on each side; compare this to Blackman or Welch windows, commonly used by other authors, which smear the tone over several adjacent bins. This is of concern for calculating SNR as we shall see in Example 1.4. □

UNCERTAINTY AND AVERAGING

The second reason why periodograms are inaccurate is as follows: the periodogram at a single frequency $P(n)$ is an estimate of a continuous function $\hat{P}(f)$ over a frequency range f_s/N centered at f_n. It turns out the estimate $P(n)$ has a standard deviation of 100% of the "actual" value. However, by taking K successive sets of N output bits, finding the periodogram of each, and *averaging* them, the standard deviation in each frequency bin is reduced by $\sqrt{K}$ [Pre92].

Example 1.3: Figure 1.10 is a striking illustration of the effect of averaging on the output power spectrum of a second-order $\Delta\Sigma$M. The upper-left graph shows the FFT of $N = 4096$ output bits; the upper-right graph depicts the average of $K = 4$ successive sets of N output bits. The following graphs are for $K = \{16, 256, 1024, 16384\}$; the graphs become smoother and smoother as the variance in each frequency bin is reduced. Moreover, the detail of the tones near $f_s/2$ is enhanced.

To generate the graph for $K = 16384$ we must calculate $N \times K \approx 67 \times 10^6$ output bits, and that takes about 12 minutes with a C program on an unloaded 170MHz Sparc Ultra. We do not usually need that large a K; it was provided merely as an illustration. 256 or fewer would suffice for most purposes. □

Matlab's `spectrum` function has parameters which control both the type of window and the number of periodogram averages taken.

3.2 SIGNAL TO NOISE RATIO (SNR)

One of the most important performance measures of a $\Delta\Sigma$M is its signal-to-noise ratio (SNR). From this we may calculate other important performance measures such as its dynamic range (DR) and peak SNR (SNR_{max}).

To find the SNR in a Nyquist-rate converter, we would divide the signal amplitude by the integrated noise from 0 to $f_N/2$ [Kes90a], where $f_N = f_s$ and hence $f_N/2 = f_s/2$. A $\Delta\Sigma$M is an oversampled converter, however, so

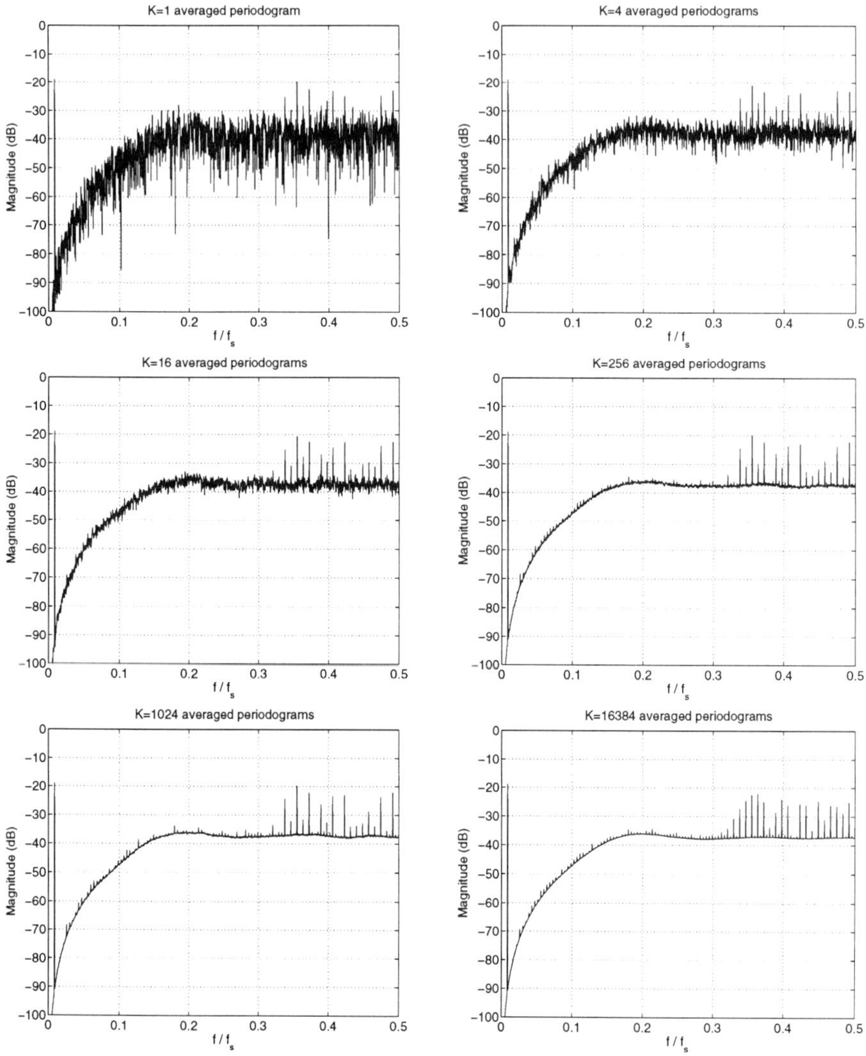

Figure 1.10. Effect of averaging on spectrum variance.

we do the same calculation over the bandwidth from 0 to $f_N/2$, which is now $f_s/(2 \cdot \mathrm{OSR})$. As noted earlier, this assumption is the same as having the modulator followed by a brick-wall low pass filter which cuts off sharply at $f_N/2$. That being said, we are about to see that the calculation is a bit more subtle than might be first imagined.

Example 1.4: Consider a 4096-point simulation of a second-order modulator. With $K = 256$ averaged periodograms, the spectrum near dc

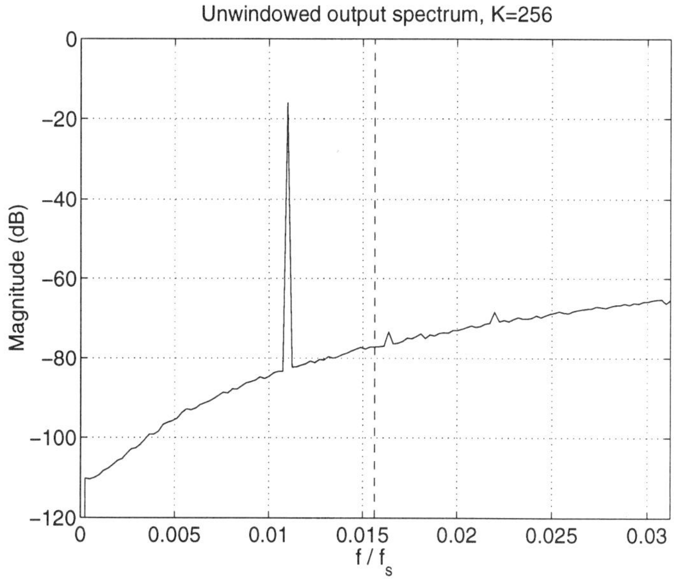

Figure 1.11. Unwindowed averaged periodogram near dc.

appears as in Figure 1.11. The input tone is -13dB and it occurs in bin $b = 45$, which is $0.01099 f_s$. Let us try to calculate the SNR for OSR $= 32$.

We must integrate the noise between 0 and $f_s/64$, which is shown by the dashed line in Figure 1.11. This corresponds to bin numbers 0 through $4096/64 = 64$. Presumably, the noise power we're interested in can be found from

$$P_n = \sum_{i=0}^{64} P(i) - P(b). \tag{1.9}$$

However, do we include bin 64 in the calculation, or exclude it? In other words, should we find the noise for $0 \leq f \leq f_s/64$, or $0 \leq f < f_s/64$? How much difference does it really make? Moreover, what should we do about the bin containing the signal? Do we subtract it as in (1.9) and leave it at that, or perhaps do something to make up for the missing bin like adding the geometric mean of the power in the surrounding bins to P_n?

Table 1.1 shows what happens with various calculations, and also shows the effect of K (the number of averaged periodograms) on the calculated SNR. For ten different runs at each K value, the SNR was calculated by dividing $P(b)$ by the quantity listed at the top of each table column and taking $10 \log_{10}$ of the result. The table lists the average and standard deviation σ of the ten SNR values, all in dB. First, we note that including bin 64

Table 1.1. Comparison of SNR calculation methods. Including the bin at $f_s/(2 \cdot \mathrm{OSR})$ lowers SNR by 0.3dB, while trying to account for the tone bin lowers it further by 0.1dB.

K	$\sum_{i=0}^{63} P(i) - P(b)$	$\sum_{i=0}^{64} P(i) - P(b)$	$\sum_{i=0}^{64} P(i) - P(b) + \sqrt{P(b-1)P(b+1)}$
1	$50.86,\ \sigma = 1.01$	$50.53,\ \sigma = 0.85$	$50.46,\ \sigma = 0.86$
4	$49.81,\ \sigma = 0.62$	$49.49,\ \sigma = 0.62$	$49.40,\ \sigma = 0.60$
16	$50.03,\ \sigma = 0.32$	$49.64,\ \sigma = 0.28$	$49.56,\ \sigma = 0.28$
64	$49.87,\ \sigma = 0.22$	$49.55,\ \sigma = 0.19$	$49.47,\ \sigma = 0.19$
256	$49.93,\ \sigma = 0.22$	$49.60,\ \sigma = 0.22$	$49.52,\ \sigma = 0.22$

lowers SNR by 0.3dB or so, while adding the geometric mean of the bins around the tone makes another 0.1dB of difference. Second, we note that σ is higher for small K—that is, the variance in calculated SNR between different runs is greater when we do less averaging. Third, calculated SNR drops by a full dB between $K = 1$ and $K = 256$. We see, therefore, that the calculation method can make more than a trivial difference to the result.

On top of this is the confusion about what happens when we window the periodograms. Figure 1.12a is another run with $K = 256$, but now a Hann window is applied to the data before finding its spectrum; the dotted line is the data from Figure 1.11 reproduced for reference. In Figure 1.11, the tone was only in one bin, and its power was $P(b) = -15.99$dB. Now, we find the tone spreads over three bins, and $P(b-1)+P(b)+P(b+1) = -20.25$dB. The unwindowed SNR for bins 0 to 64 excluding bin b was 49.66dB; the Hann-windowed SNR for bins 0 to 64 excluding bins $b-1$ to $b+1$ is 50.57dB.

We would like to understand two things: why the tone power changed, and why the SNR changed. The difference of -4.26dB in tone power can be explained as follows. The periodogram of (1.7) is normalized such that the signal power in time and frequency are equal (i.e., Parseval's theorem holds). Since the output sequence is composed of ± 1, the power in time is 1; we can easily verify that $\sum_{i=0}^{N/2+1} P(i) = 1$ in Matlab. A Hann window turns out to scale the total power by 0.375, and $10\log_{10}(0.375) = -4.2597$— exactly the difference seen in the tone power. The total baseband noise seems to have been reduced by $50.57 - 49.66 + 4.26 = 5.15$dB. It is not so easy to explain numerically where the extra $5.15 - 4.26 = 0.91$dB of noise reduction by windowing comes from, though qualitatively we expect the reduction because windowing is known to reduce leakage problems.

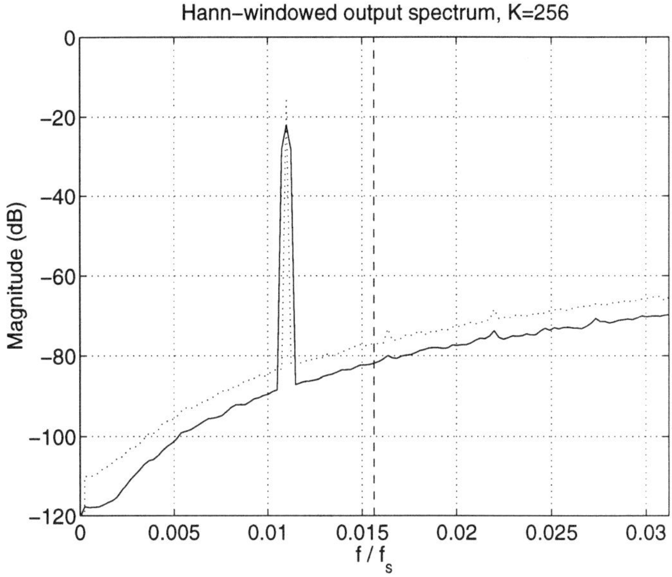

Figure 1.12a. Hann-windowed averaged periodogram near dc.

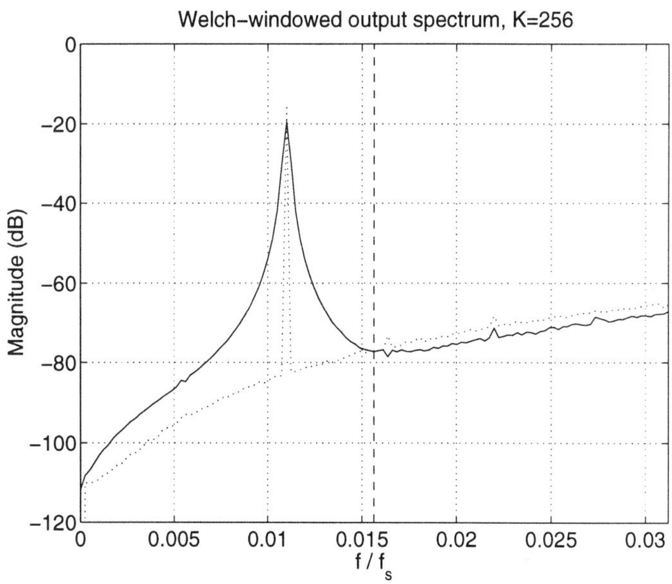

Figure 1.12b. Welch-windowed averaged periodogram near dc.

Finally, as alluded to in Example 1.2, we vastly prefer a Hann window for SNR calculations over other windows. Figure 1.12b shows what a Welch window does to the baseband spectrum with the unwindowed spectrum plotted for reference. The tone has been smeared over so many bins that it becomes impossible to know where the noise begins. We only have 64 bins in which to find the noise, and too many of them get corrupted by smearing for a meaningful SNR calculation[2]. □

The preceding example illustrates that SNR can vary by about 1dB depending on how the calculation is done. This suggests that specifying SNR to more than one decimal place is pointless, and even the first decimal place might not be very meaningful. Unfortunately, the example does little to clear up confusion about the "right" way to calculate SNR; papers in the literature rarely seem to be specific enough to tell what they do, and even if they did, there'd *still* be disagreement. We arbitrarily adopt the definition in the first column of Table 1.1, where we simply neglect the tone bin(s) and the final FFT bin.

Some authors refer to signal-to-quantization-noise ratio (SQNR), where only quantization noise power is counted as noise, as distinct from signal-to-noise-and-distortion ratio (SNDR or SINAD), where both quantization noise power *and* the power in any output harmonics of the input signal are counted. We use SNR to mean SNDR—our SNR calculations will include any power in harmonics of the input signal caused by distortion. We shall examine some of the things that can create input signal harmonics in the output spectrum presently.

3.3 OTHER PERFORMANCE MEASURES

DYNAMIC RANGE

The dynamic range of a $\Delta\Sigma$M, often specified in decibels, is equivalent to the resolution of the modulator as an ADC. We can convert from resolution in dB to resolution in bits by relating a $\Delta\Sigma$M to a Nyquist-rate converter using [Ben48]

$$\text{DR(bits)} = (\text{DR(dB)} - 1.76)/6.02. \tag{1.10}$$

To actually find the DR for a given modulator, SNR is plotted against input amplitude. The input amplitude range which gives SNR ≥ 0 is precisely the DR.

Example 1.5: For a second order low pass $\Delta\Sigma$M, Figure 1.13a shows the SNR as a function of input amplitude for two different OSRs, 32 and 64. We call this kind of graph a *dynamic range plot*. The slope of each curve

[2]It might be worth noting that just about every author has their own pet window; in the final analysis, it probably makes little difference which window is chosen.

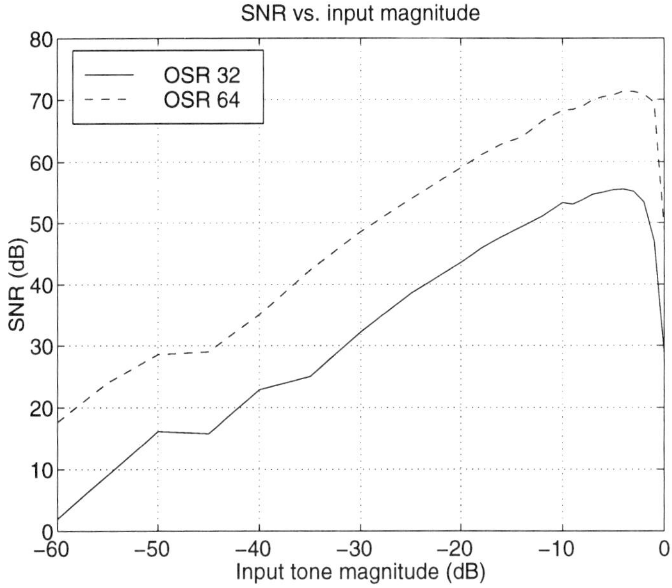

Figure 1.13a. Dynamic range plot for ideal double integration ΔΣM.

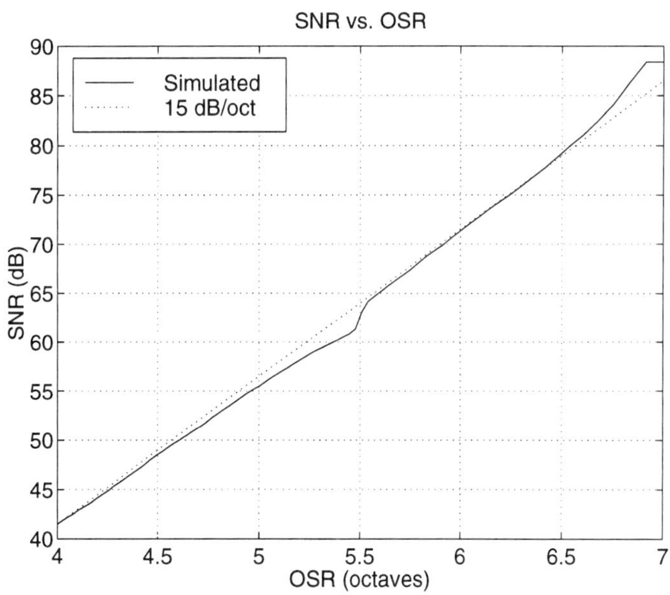

Figure 1.13b. Improvement of double integration ΔΣM SNR with oversampling.

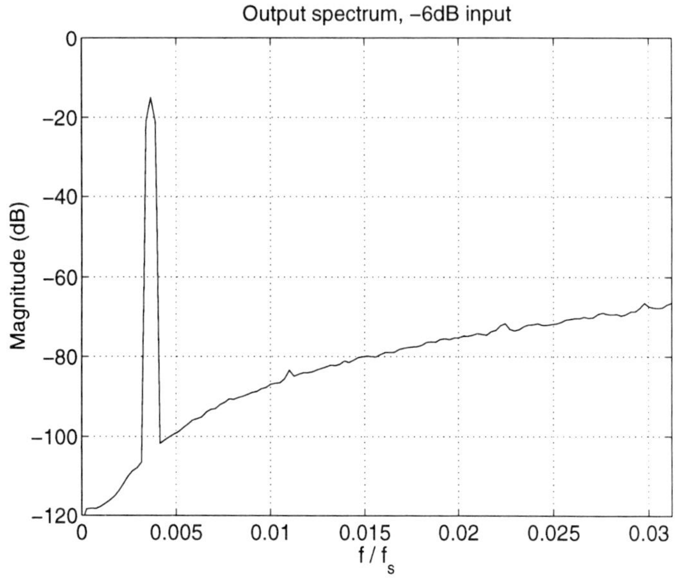

Figure 1.14a. Ideal double integration ΔΣM spectrum with −6dB input.

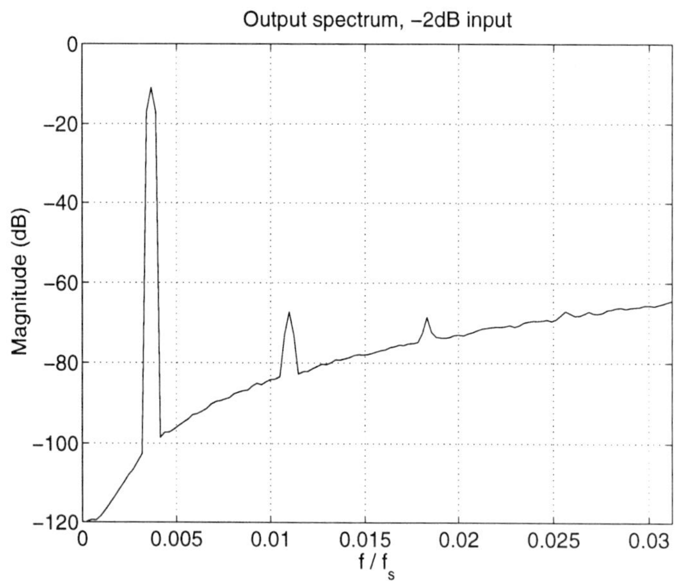

Figure 1.14b. Ideal double integration ΔΣM spectrum with −2dB input.

is 1dB/dB except for large input amplitudes where the SNR stays constant or decreases with input amplitude. For small inputs, the SNR is limited by the in-band noise, while at large inputs, the SNR starts to become affected by input signal harmonics. Figures 1.14a and b show the baseband output spectrum for inputs of -6dB and -2dB, respectively. Signal harmonics are clearly present for the larger input.

Extrapolating to SNR $=$ 0 for small inputs indicates the DR for OSR $= 32$ is 62dB, or about 10 bits from (1.10), and for OSR $= 64$ the DR is 77dB (about 12.5 bits). We said earlier that at a fixed input amplitude SNR improves as $6m + 3$dB per octave of oversampling, where m is the order of the modulator. Figure 1.13b demonstrates the truth of this for a -4dB input tone. □

FULL SCALE AMPLITUDE

In the previous example we referred to the input as being in dB, but what we did not make explicit is that it is dB relative to full scale[3]. How is "full scale" defined for a $\Delta\Sigma$M? The answer is not always obvious. In many cases, a full-scale input is one whose magnitude equals the maximum magnitude of the quantizer feedback, assuming a quantizer whose output is centered at 0 (which it almost always is). For an input larger than this, the feedback will not be able to keep the modulator stable; we refer to this as *overloading the modulator*.

Example 1.6: In the previous example, the quantizer was feeding back ± 1. When the input was a tone with peak amplitude 0.1V, it transpired that the tone appeared in the output spectrum with magnitude -23.01dB, an rms value which corresponds to a peak value of -20dB $= 0.1$V. We can deduce that 1V is the full-scale input level in that example. An input larger than 1V will overload the modulator. Inputs close to 1V cause graceful degradation of SNR due to increased spectral harmonic content, as we saw in Figure 1.14a. □

Example 1.7: Figure 1.15 shows a typical implementation of a second-order low pass CT $\Delta\Sigma$M for high-speed ADC. The input signal is fed through a transconductor g_{m1}, and at the transconductor output node, there is a feedback current of magnitude k_2. The current $g_{m1}u$ can be no larger than k_2 without overloading the modulator; therefore, the full-scale input signal magnitude is k_2/g_{m1}. Typical component values might be $k_2 = 0.4$mA and g_{m1} $=$ 1mA/V, so a 0.4V input signal would appear at the output as 0dB when the output bits are ± 1. □

[3]It would probably be less confusing if the units of the input signal were explicitly specified as "dBrel" or something similar to indicate that it is dB relative to some maximum. However, most of the literature refers to "dB", so we do the same here.

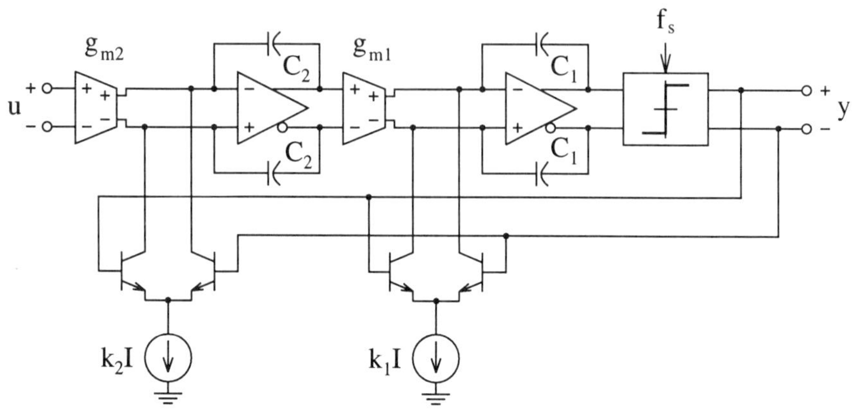

Figure 1.15. InP second-order CT ΔΣM by Jensen et al.

For certain more complicated modulator structures we will encounter later, the full-scale input range will need to be found from simulation rather than calculation.

MAXIMUM SNR AND MAXIMUM STABLE AMPLITUDE

Maximum SNR, SNR_{max} (sometimes called "peak SNR"), is easily found from a DR plot as the peak of the SNR vs. input amplitude curve. It turns out that the second-order low pass ΔΣM is stable all the way up to an input amplitude of 0dB [Wan92]. It also turns out that higher-order modulators usually become unstable before 0dB is reached; this instability usually manifests itself in clipping of the final integrator output which causes the quantizer to produce a long consecutive sequence of the same output bit. This means the signal encoding properties of the modulator become poor [Ris94] and hence SNR is degraded. The maximum stable amplitude (MSA) is, then, the largest input amplitude which keeps the final integrator output bounded most of the time. It, too, can be found from a DR plot as the maximum input amplitude for which $SNR \geq 0$.

SPURIOUS FREE DYNAMIC RANGE

Nyquist-rate ADCs sometimes specify a rating for spurious free dynamic range (SFDR) [Kes90b]. To measure SFDR, we apply a tone at the ADC input and look for the largest *spur* between 0 and $f_N/2$, where a spur is a tone visible above the noise floor. In theory, we must do this for all input frequencies and phases to find the very worst-case spur. Then, SFDR is the largest magnitude difference between the amplitudes of the input tone and the largest spur in dB, over all input tone amplitudes.

The importance of SFDR depends on the application. In some applications, a good SFDR is more important than a good DR. In radio systems, for example, it might be important to keep the amplitude of spurious tones low since nonlinearities might cause them to intermodulate and corrupt the desired signal, while the total amount of in-band noise might not matter so much. SFDR measurements are only rarely quoted for oversampling converters such as $\Delta\Sigma$Ms [Jen95]. Realistically, an SFDR measurement can only be performed on an actual circuit rather than in simulation because it requires many different input amplitudes, frequencies, and phases. We will usually neglect SFDR in our examination of CT $\Delta\Sigma$Ms until we come to Chapter 7 where we explicitly measure it for a fabricated design.

4. SIMULATION METHODS

To characterize the performance of a $\Delta\Sigma$M, we take the spectrum of its output bit stream. How do we actually generate this output bit stream in a simulation? Determining the output bits analytically is very difficult because of the nonlinear quantizer. As a result, time-domain simulation of the modulator is the usual method. In the simulation of just about any system, there exists a tradeoff between *realism* and *simulation time*: as we model the behavior of a system more accurately, the length of time required to generate simulation results increases. Let us first consider our simulation options for DT $\Delta\Sigma$Ms, a subject which has received a considerable amount of attention in the literature, followed by those for CT $\Delta\Sigma$Ms [Che98a].

4.1 DISCRETE-TIME MODULATOR SIMULATION

An ideal DT $\Delta\Sigma$M can be described by a discrete-time system of equations. For the general modulator in Figure 1.16 which includes input prefiltering

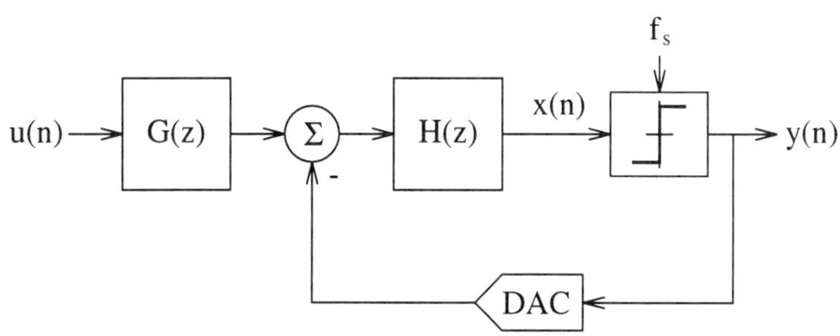

Figure 1.16. A general DT $\Delta\Sigma$M including input prefiltering.

[Ris94], we can write a linear equation for the quantizer input in terms of the circuit input and quantizer output

$$X(z) = G(z)H(z)U(z) - H(z)Y(z). \tag{1.11}$$

$G(z)$ and $H(z)$ are rational functions of z, with $G(z)$ proper and $H(z)$ strictly proper. It is a trivial matter to take the inverse $\mathcal{Z}$-transform of (1.11), which leaves an expression for $x(n)$, the quantizer input now, in terms of past samples of (u, x, y):

$$x(n) = \sum_{k=1}^{m} a_k x(n-k) + \sum_{k=1}^{m} b_k u(n-k) + \sum_{k=1}^{m} c_k y(n-k). \tag{1.12}$$

$\{a_k, b_k, c_k\}$ are constants that can be found from $G(z)$ and $H(z)$. For each $x(n)$ found from (1.12), we find $y(n)$ by assuming an ideal quantizer; in the case of a single-bit quantizer,

$$y(n) = \begin{cases} +1, & x(n) \geq 0 \\ -1, & x(n) < 0. \end{cases} \tag{1.13}$$

Applying (1.12) and (1.13) for $n = 1, \ldots, N$ in a high-level language such as Matlab or C gives a very rapid method for determining the output bit stream.

Rapidity is one thing, but realism is another. A practical circuit will likely not be represented by its ideal equations. Eventually, we will have a transistor-level description of a circuit whose behavior we would like to simulate, and it is most likely that we will turn to a full-circuit simulator such as SPICE or Spectre. While it is probably that such a simulation is able to model most, if not all, of the pertinent nonidealities which affect circuit performance, we will often be stuck waiting for hours or even days while generating enough output bits for an FFT. A detailed discussion of these nonidealities appears in Chapter 3.

Fortunately, there exists more than one "middle-ground" approach, where reasonable accuracy is achieved while maintaining acceptably-fast simulation speed. Several programs (Simulink under Matlab [Sim96], SPW [SPW92], and Ptolemy [Pto97] among them) allow a system to be defined at the block diagram level graphically, with the function of each block controlled by the user. This allows both rapid, user-friendly prototyping of ΔΣM systems along with the inclusion of nonidealities (such as finite integrator output swing and quantizer hysteresis) by using the appropriate blocks in the simulation. In a similar manner, full-circuit simulation programs like SPICE and Spectre[4] often allow the specification of a circuit with *macromodels*, where a block is modeled as

[4]Spectre is perhaps more suited to discrete-time block-level simulation than SPICE.

an ideal version of itself instead of as a transistor-level description. Better still, ideal blocks can be replaced one at a time with transistor-level descriptions, which allows the user to see the effect of nonidealities in each individual block on overall modulator performance while keeping the simulation speed faster than for a complete transistor-level circuit.

Best of all, there exist special-purpose programs written specifically for the simulation of DT ΔΣMs. Both MIDAS [Wil92] and TOSCA [Lib93] are examples of programs which can simulate and extract key performance parameters from otherwise ideal ΔΣMs as well as DT ΔΣMs which include important nonidealities such as finite op amp gain, finite switch on-resistance, and clock feedthrough. A program by Medeiro et al. [Med95] goes even further: the user specifies modulator parameters such as required resolution, clock rate, and power consumption, and then the program can design and automatically produce the circuit layout for a complete SC modulator which meets the specifications.

Clearly, a first-time DT ΔΣM designer has plenty of options for generating an output bit sequence relatively quickly while still including the effects of relevant nonidealities.

4.2 CONTINUOUS-TIME MODULATOR SIMULATION

The situation for CT ΔΣMs is perhaps not as good, most likely because there has been considerably less attention devoted to the design of CT ΔΣMs. Nonetheless, there are several choices. As with DT ΔΣMs, we may represent an ideal CT ΔΣM with a frequency domain equation akin to (1.11),

$$\hat{X}(s) = \hat{G}(s)\hat{H}(s)\hat{U}(s) - \hat{H}(s)\hat{Y}(s), \tag{1.14}$$

where we use the continuous frequency variable s rather than the discrete one $z = \exp(sT_s)$ and T_s is the sampling frequency. Taking the inverse Laplace transform of (1.14) does not lead to as easily-implementable an equation as that which resulted from the discrete case (1.12), so for time-domain simulation we must represent the system as a series of coupled first-order differential equations and solve them using numerical integration.

Example 1.8: For the modulator in Figure 1.15, the equations describing circuit behavior are

$$\begin{aligned} C_1 \frac{d\hat{x}_1}{dt} &= g_{m1}\hat{u}(t) + k_2\hat{y}(t) \\ C_2 \frac{d\hat{x}_2}{dt} &= g_{m2}\hat{x}_1(t) + k_1\hat{y}(t) \end{aligned} \tag{1.15}$$

and the single-bit quantizer is described by (1.13). Implementing these equations in a numerical integration program is slightly more tedious than solving the difference equations, but not terribly difficult to do. □

It happens that because of the clocked quantizer inside the CT loop, an ideal CT modulator has a DT equivalent. Thus, there exists a mapping between the s-domain description of a $\Delta\Sigma$M and the z-domain which can be exploited to increase simulation speed and give intuitive understanding of modulator behavior. We discuss this further in the next chapter.

Once again, ideal CT $\Delta\Sigma$M behavior is one thing and the behavior of a real circuit is another. As with DT modulators, full-circuit simulation of CT modulators is painfully slow when each block is described down to the transistor level. Also as in the DT case, macromodel simulation is an attractive option for reducing simulation time while incorporating key nonidealities: a CT $\Delta\Sigma$M is first described with ideal blocks in a full-circuit simulator, then nonidealities can be added gradually to observe the effect on performance. Often, graphical block diagram simulators (such as those listed in the previous section) can also simulate CT systems, so this too is a choice for CT $\Delta\Sigma$Ms.

As far as specialized CT $\Delta\Sigma$M simulation tools go, the literature seems not to mention large-scale efforts. Frequently [Bro90, Cha92, Ush94, Che98a] special-purpose programs in a high-level language such as C are written in the course of studying modulator performance. Opal [Opa96] has built a fairly general framework based on the CT/DT equivalence mentioned above, but to our knowledge there is no CT $\Delta\Sigma$M equivalent of a program like TOSCA.

Rapid and realistic simulation of CT $\Delta\Sigma$Ms is a central underlying theme of this monograph. We will be making use of various simulation techniques as we delve into detail, and we will describe them as they arise. To summarize our general approach quickly, we use a custom C program combined with behavioral modeling.

5. SUMMARY

Delta–sigma modulation is a technique which combines filtering and oversampling to perform analog-to-digital conversion: the noise from a low resolution quantizer is shaped away from the signal band prior to being removed by filtering. High-speed conversion can be accomplished by using a continuous-time filter inside the delta–sigma loop, and we are interested in this for its applicability to radio receiver and other high frequency circuits. Continuous-time modulators have other advantages such as built-in antialiasing and less switching noise. Performance of a $\Delta\Sigma$M is determined by taking the spectrum of a sequence of output bits generated from time-domain simulation of the modulator; it is characterized with some of the usual ADC performance measures such as DR and SNR, while omitting others which have no meaning in $\Delta\Sigma$Ms such as DNL and INL. How to actually perform the time-domain simulation is a matter of considerable import in oversampled converters because they will usually require many more output samples than a Nyquist rate ADC before performance can be measured.

Chapter 2

DESIGNING CONTINUOUS-TIME MODULATORS

This chapter explains how to design continuous-time modulators. We explain and illustrate a transformation that allows us to design CT modulators in the DT domain, and then present some techniques that can be used in SPICE or another circuit simulator to verify the correctness of practical designs.

1. IDEAL $\Delta\Sigma$M DESIGN

There is a great deal of literature and software available to help us choose a DT $\Delta\Sigma$M loop filter $H(z)$ that will meet given a given resolution requirement. We show in this section that designing a CT $\Delta\Sigma$M is almost no extra work: our recommended design procedure is to start by picking an $H(z)$, then we transform it to the *equivalent* $\hat{H}(s)$. We illustrate one of the nice features of CT $\Delta\Sigma$Ms: they have some antialiasing built right into them. Finally, we present some design examples to clarify the use of the transformation method.

Schreier presents a state-space method of designing ideal CT $\Delta\Sigma$Ms in [Nor97, Chap. 6]. We prefer to work with the loop filter in pole-zero form, though either method works.

1.1 THE IMPULSE-INVARIANT TRANSFORM

We claim above that a DT modulator has a CT equivalent which can be found through a transformation between the DT and CT domains. The reader might be familiar with the existence of $s \leftrightarrow z$ mappings: in switched-capacitor filter design, for example, we have the forward Euler, backward Euler, and bilinear transforms [Gre86, Chap. 2]. It turns out that we desire the *impulse responses* of the CT and DT $\Delta\Sigma$M loops to be the same. Let us explain why.

First, what condition must be satisfied for two modulators to be equivalent? Suppose we have two modulators to which we apply the same input waveform.

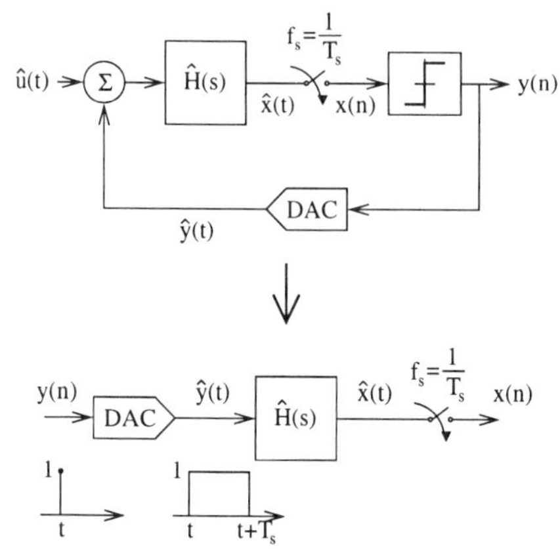

Figure 2.1a. Open loop CT $\Delta\Sigma$M.

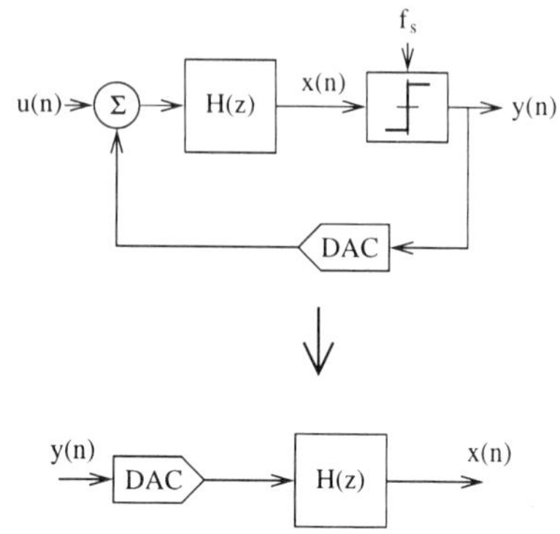

Figure 2.1b. Open loop DT $\Delta\Sigma$M.

If we simulate each modulator in the time domain, and we find they both produce the same sequence of output bits, then they're likely to be equivalent modulators. We can *guarantee* they'll produce the same output bit sequence if we ensure that the inputs to their *quantizers* are the same at sampling instants—at a given instant, each quantizer would then make the same decision about what output bit to produce, and thus the same bits would be combined with the (identical) input to produce the same quantizer input voltage at the *next* sampling instant. Therefore, we can see that *two modulators are equivalent if, for the same input waveform, their quantizer input voltages at sampling instants are equal.*

To see how to make use of this in CT modulator design, we take the CT $\Delta\Sigma$M at the top of Figure 2.1a. Rather than making the sampler implicitly inside the quantizer, we have chosen to make it explicit prior to the quantizer—this does not change modulator behavior. In the bottom part of Figure 2.1a, we have both opened the loop around the quantizer and zeroed the modulator input; the rationale for this will be explained shortly. When we do the same thing to a DT $\Delta\Sigma$M, Figure 2.1b, we have two open-loop diagrams (a CT one and a DT one) to compare.

The input to the CT open-loop diagram is the output bit $y(n)$. This is a discrete-time quantity—it only ever changes at sampling instants. The DAC may be thought of as a *discrete-to-continuous converter*: it takes a sample $y(n)$ and produces some kind of continuous pulse $\hat{y}(t)$, depicted in Figure 2.1a as a full-period rectangular pulse. This pulse is filtered by the loop filter $\hat{H}(s)$, then sampled, where it becomes the discrete-time output of the open-loop system

$x(n)$ (as well as being the input to the quantizer). In the DT open-loop diagram, the signals never leave the DT domain.

Note that for both CT and DT loops, the input and output of the open-loop diagrams are both DT quantities. We just said that two $\Delta\Sigma$Ms are equivalent when their quantizer inputs are the same at sampling instants; for the CT and DT modulators, this means

$$x(n) = \hat{x}(t)|_{t=nT_s}. \tag{2.1}$$

This would be satisfied if the impulse responses of the open-loop diagrams in Figures 2.1a and b were equal at sampling times. This leads to the condition [Thu91]

$$\mathcal{Z}^{-1}\{H(z)\} = \mathcal{L}^{-1}\{\hat{R}_D(s)\hat{H}(s)\}|_{t=nT_s}, \tag{2.2}$$

or, in the time domain [Sho96],

$$h(n) = [\hat{r}_D(t) * \hat{h}(t)]|_{t=nT_s} = \int_{-\infty}^{\infty} \hat{r}_D(\tau)\hat{h}(t-\tau)d\tau|_{t=nT_s} \tag{2.3}$$

where $\hat{r}_D(t)$ is the impulse response of the DAC. This transformation between CT and DT is called the *impulse-invariant transformation* [Gar86] because we require the open-loop impulse responses to be the same at sampling instants. (It was first proposed by F. M. Gardner of phase-locked loop fame.)

The combination of removing the quantizer and zeroing the input allows us to study the behavior of just the linear portion of the loop on the output bit (which includes the quantization noise). Put another way, it allows us to see how to make the NTFs of the CT and DT modulators the same. Thus, suppose we have a DT modulator with a filter $H(z)$ that gives us particular noise-shaping behavior. We can build a CT modulator with identical noise-shaping behavior by first choosing a DAC pulse shape $\hat{r}_D(t)$, and then using either (2.2) or (2.3) to find $\hat{H}(s)$. This is immensely powerful: there is a good deal of literature and software[1] available for choosing $H(z)$, all of which we can use before the final (and relatively easy) step of transforming $H(z)$ to $\hat{H}(s)$.

Without loss of generality, we normalize the sampling period to $T_s = 1$ second to simplify the following discussion.

1.2 EXPLOITING THE EQUIVALENCE

To actually do the transform, we proceed as follows. First, we write $H(z)$ as a partial fraction expansion. Then, we choose a DAC pulse shape; for simplicity, we assume a perfectly rectangular DAC pulse of magnitude 1 that

[1]We highly recommend Richard Schreier's Delta–Sigma Toolbox for Matlab, available over the Internet at `ftp://next242.ece.orst.edu/pub/delsig.tar.Z`.

lasts from α to β, i.e.,

$$\hat{r}_{(\alpha,\beta)}(t) = \begin{cases} 1, & \alpha \le t < \beta,\ 0 \le \alpha < \beta \le 1 \\ 0, & \text{otherwise.} \end{cases} \tag{2.4}$$

This covers most types of practical DAC pulse. Finally, we use Table 2.1 to convert each partial fraction pole from z to s, and recombine the results to get $\hat{H}(s)$. The table was found by solving (2.2) in the symbolic math program Maple [Red94] where the Laplace transform of (2.4) is

$$\hat{R}_{(\alpha,\beta)}(s) = \frac{\exp(-\alpha s) - \exp(-\beta s)}{s}. \tag{2.5}$$

What we find is, a z-domain pole of multiplicity l at z_k maps to one at s_k with the same multiplicity, with

$$s_k = \ln z_k \tag{2.6}$$

(this makes sense when you think of $z = \exp(sT_s)$). Poles at dc (i.e., $z_k = 1$) end up giving $0^l/0^l$ as the numerator of the s-domain equivalent, which necessitates l applications of l'Hôpital's rule; this has been done in the right column of Table 2.1.

Example 2.1: Many designs use DACs with an output pulse which remains constant over a full period, which we shall term a *non-return-to-zero* (NRZ) DAC. For this type of DAC, $(\alpha, \beta) = (0, 1)$ in (2.4). Moreover, we shall see that many high-speed designs are second-order low pass designs; these differentiate the quantization noise twice so that $\text{NTF}(z) = (z-1)^2$ and

$$H(z) = \frac{-2z+1}{(z-1)^2}. \tag{2.7}$$

Writing this in partial fractions yields

$$H(z) = \frac{-2}{z-1} + \frac{-1}{(z-1)^2}. \tag{2.8}$$

Thus $z_k = 1$, which means $s_k = 0$ from (2.6). Applying the first row of Table 2.1 to the first term of (2.8) and the second row to the second term with $(\alpha, \beta) = (0, 1)$ gives

$$\hat{H}(s) = \frac{-2}{s} + \frac{-1+0.5s}{s^2} \tag{2.9}$$

$$= -\frac{1+1.5s}{s^2}. \tag{2.10}$$

(2.10) was first derived by Candy [Can85] as the CT equivalent of the DT double integration modulator in (2.7).

Table 2.1. s-domain equivalences for z-domain loop filter poles.

z-domain pole	*s-domain equivalent*	*Limit for* $z_k = 1$
$\frac{y_0}{z-z_k}$	$\frac{r_0}{s-s_k} \times \frac{y_0}{z_k^{1-\alpha}-z_k^{1-\beta}}$ $r_0 = s_k$	$\frac{r_0}{s-s_k}$ $r_0 = \frac{y_0}{\beta-\alpha}$
$\frac{y_0}{(z-z_k)^2}$	$\frac{r_1 s+r_0}{(s-s_k)^2} \times \frac{y_0}{z_k(z_k^{1-\alpha}-z_k^{1-\beta})^2}$ $r_1 = q_1 s_k + q_0$ $r_0 = q_1 s_k^2$ $q_1 = z_k^{1-\beta}(1-\beta) - z_k^{1-\alpha}(1-\alpha)$ $q_0 = z_k^{1-\alpha} - z_k^{1-\beta}$	$\frac{r_1 s+r_0}{(s-s_k)^2}$ $r_1 = \frac{1}{2}\frac{(\alpha+\beta-2)y_0}{\beta-\alpha}$ $r_0 = \frac{y_0}{\beta-\alpha}$
$\frac{y_0}{(z-z_k)^3}$	$\frac{r_2 s^2+r_1 s+r_0}{(s-s_k)^3} \times \frac{y_0}{z_k^2(z_k^{1-\alpha}-z_k^{1-\beta})^3}$ $r_2 = \frac{1}{2}q_2 s_k - q_1$ $r_1 = -q_2 s_k^2 + q_1 s_k + q_0$ $r_0 = \frac{1}{2}q_2 s_k^3$ $q_2 = (1-\beta)(2-\beta)(z_k^{1-\beta})^2$ $+ (1-\alpha)(2-\alpha)(z_k^{1-\alpha})^2$ $+ [\beta(\beta+3)+\alpha(\alpha+3)$ $- 4(1+\alpha\beta)]z_k^{1-\alpha} z_k^{1-\beta}$ $q_1 = (\frac{3}{2}-\beta)(z_k^{1-\beta})^2$ $+ (\frac{3}{2}-\alpha)(z_k^{1-\alpha})^2$ $+ (\alpha+\beta-3)z_k^{1-\alpha} z_k^{1-\beta}$ $q_0 = (z_k^{1-\alpha} - z_k^{1-\beta})^2$	$\frac{r_2 s^2+r_1 s+r_0}{(s-s_k)^3}$ $r_2 = \frac{1}{12}\frac{y_0}{\beta-\alpha}[\beta(\beta-9)$ $+ \alpha(\alpha-9)+4\alpha\beta+12]$ $r_1 = \frac{1}{2}\frac{(\alpha+\beta-3)y_0}{\beta-\alpha}$ $r_0 = \frac{y_0}{\beta-\alpha}$

Just so we state it here, to remove the $T_s = 1$ normalization, replace every s in $\hat{H}(s)$ with sT_s. Therefore,

$$\hat{H}(s) = -\frac{1 + 1.5sT_s}{(sT_s)^2} \tag{2.11}$$

is the more general way to write the loop filter for the double integration NRZ modulator. □

Example 2.2: We shall see that there are some advantages to using a return-to-zero (RZ) DAC in low pass modulators. These DACs produce a rectangular pulse which lasts from 0 to $T_s/2$ only. To find the CT equivalent for the double integration modulator in (2.7), we apply Table 2.1 to the terms of (2.8) with $(\alpha, \beta) = (0, 0.5)$ to get

$$\begin{aligned} \hat{H}(s) &= \frac{-4}{s} + \frac{-2 + 1.5s}{s^2} && (2.12) \\ &= -\frac{2 + 2.5s}{s^2}. && (2.13) \end{aligned}$$

The numerator coefficients are larger than (2.10), which makes sense intuitively because in order to get the same quantizer input voltage with a shorter DAC pulse, we require larger integrator gains. □

Example 2.3: Matlab has some built-in functions to do these transformations, at least in the NRZ DAC pulse case. To transform the double integration modulator, we type

```
[Hsnum,Hsden] = d2cm([0 -2 1],[1 -2 1],1)
```

`d2cm` is the discrete-to-continuous conversion routine of the control toolbox. The first two arguments are the numerator and denominator of $H(z)$ in descending powers of z (so $-2z + 1$ in (2.7) becomes `[0 -2 1]` while $(z-1)^2$ is expanded to `[1 -2 1]`), while the third argument is the sample period T_s. Matlab returns

```
Hsnum =              0 -1.5000e+00 -1.0000e+00
Hsden = 1.0000e+00  1.3323e-15  4.4373e-31,
```

which are the numerator and denominator of $\hat{H}(s)$ in descending powers of s—in other words, $\hat{H}(s) = (-1.5s - 1)/s^2$ as in (2.10).

In principle, it is possible to extend this to other (rectangular) DAC pulse shapes. The more general form of `d2cm` under Matlab V looks like

```
[a,b,c,d] = tf2ss([0 -2 1],[1 -2 1],1)
sysd = ss(a,b,c,d,1)    % State-space, Ts=1
sysc = d2c(sysd)
```

```
tf(sysc)
```

—we convert the pole-zero form to state-space. The individual state-space matrices and vectors can be accessed using `sysc.a`, etc. [Nor97, Chap. 4] says that for general rectangular DAC pulses (α, β), only the $\hat{B}$ state-space matrix changes. It is found from the NRZ $\hat{B} \equiv \hat{B}_{(0,1)} \equiv$ `sysc.b` by applying the conversion

$$\hat{B}_{(\alpha,\beta)} = [\exp(\hat{A}(1-\alpha)) - \exp(\hat{A}(1-\beta))]^{-1}(A - I)\hat{B}_{(0,1)}. \quad (2.14)$$

While this works for many modulators, unfortunately it does *not* work for the double integration modulator because

$$\hat{A} = \begin{bmatrix} 1 & -1 \\ 1 & -1 \end{bmatrix}$$

is singular, and hence the expression involving $\exp(\hat{A})$ in (2.14) is also singular. `d2c` avoids singular matrix problems through a clever algorithm. We *can* get close enough for all practical purposes to the right answer by making $\hat{A}$ slightly nonsingular manually; in this case, for $(\alpha, \beta) = (0, 0.5)$, after the `sysc` command above we would type

```
ac = [1 -1; 1 -1.0001]
bc = inv(expm(ac*1) - expm(ac*0.5))
     * (sysd.a - eye(2)) * sysc.b
```

This returns `[0.5000 -1.5001]` for `bc`. Then,

```
sysc.b = bc
tf(sysc)
```

returns $(-2.5s - 2)/s^2$, as we found in (2.13). □

1.3 BAND PASS MODULATORS

We saw in Figure 1.7 that band pass modulators with center frequency $f_s/4$ are potentially useful in radio receivers; indeed, several circuits [Sin95, Gao97b, Jay97, Gao98a] have been built with this application in mind. We study the ideal design of BP CT modulators here, starting with the $f_s/4$ case as a particular example.

We noted in Chapter 1 §1.2 that taking a low pass NTF(z) with a quantization noise notch at dc and performing the substitution $z^{-1} \rightarrow -z^{-2}$ gives a BP NTF(z) with a noise notch at $f_s/4$, one-quarter the sampling frequency, with double the order and identical stability properties to the LP prototype. The substitution can be applied to the loop filter $H(z)$ to yield the same result.

Applying this to the double integration modulator (2.7) gives

$$H_{LP}(z) = \frac{-2z^{-1} + z^{-2}}{(1 - z^{-1})^2} \rightarrow H_{BP}(z) = \frac{2z^{-2} + z^{-4}}{(1 + z^{-2})^2}. \tag{2.15}$$

This contains two double poles at $z_k = \pm j$; we could find the equivalent $\hat{H}_{BP}(s)$ by applying the results in Table 2.1 to a partial fraction expansion of (2.15) or by simply using Matlab.

Example 2.4: Doing this for NRZ DAC pulses yields

$$\hat{H}_{BP}(s) = \frac{-1.0354s^3 + 1.0652s^2 - 1.3210s + 4.5661}{(s^2 + (\frac{\pi}{2})^2)^2}. \tag{2.16}$$

How do we build a circuit to implement this? Historically, LP DT modulators have been built as a cascade of integrators $z^{-1}/(1 - z^{-1})$ [Cha90], and building an $f_s/4$ BP DT modulator simply requires replacing the integrator blocks directly with resonator blocks $-z^{-2}/(1 + z^{-2})$. It is likewise possible to build LP CT modulators as a cascade of integrators $1/s$—the block diagram for Figure 1.15 and (2.10) is shown in Figure 2.2. However, simply

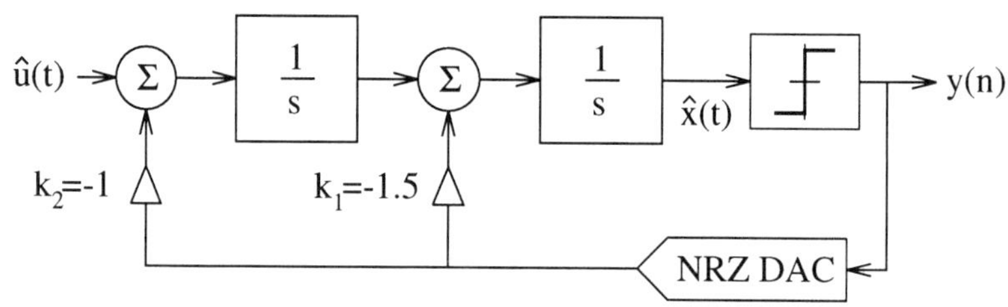

Figure 2.2. Block diagram for LP CT ΔΣM from Figure 1.15.

replacing integrators with resonators $As/(s^2 + \omega^2)$, $\omega = \pi/2$ as in Figure 2.3 does *not* build (2.16): the numerator of $\hat{H}_{BP}(s)$ for Figure 2.3 can be seen to be missing an s^0 term, yet it is required in (2.16). Moreover, the coefficients of s^3 and s^1 are both proportional to k_4, a second undesirable trait. □

The issue is one of *controllability*: we have four numerator coefficients in $\hat{H}_{BP}(s)$ and fewer than four tunable parameters (we could play with each DAC current and numerator gain separately, but not in a combination to realize the needed transfer function). Early designs [Thu91] suffered from lack of controllability. One way to get four independent tunable parameters is to use resonators with a low pass term included in the numerator: $(As+B)/(s^2+\omega^2)$. However, in a circuit implementation this might be cumbersome: it is usually easier to build purely band pass circuits.

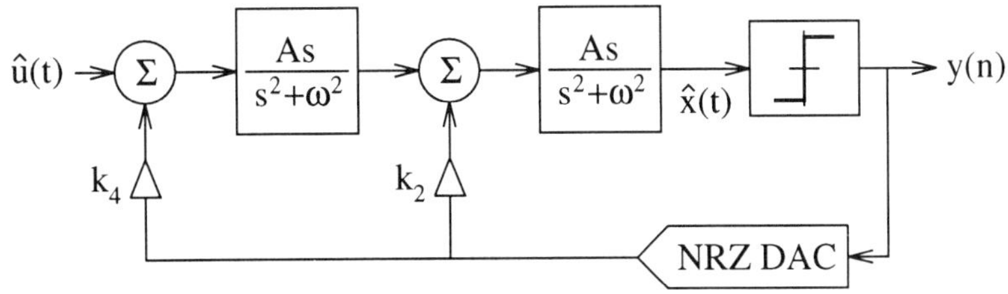

Figure 2.3. Block diagram for BP CT ΔΣM with integrators replaced by resonators that cannot implement desired equivalent $H(z)$.

Our preferred method for getting four tunable parameters is the elegant one first proposed in [Sho94] and [Sho95]: we use two identical resonators $As/(s^2 + \omega^2)$, but two *different* types of feedback DAC, each with separately tunable output currents. Full controllability is achieved through tuning the four DAC currents appropriately. This architecture has been dubbed the *multi-feedback architecture*, and an example of it appears in Figure 2.4. There,

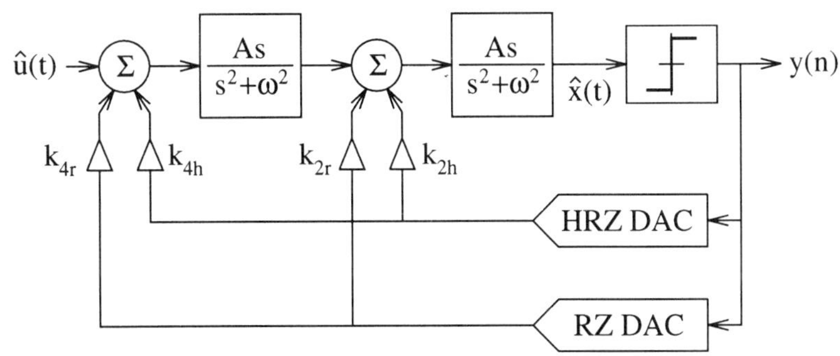

Figure 2.4. Multi-feedback BP CT ΔΣM architecture.

the DACs are return-to-zero (RZ), which has $(\alpha, \beta) = (0, 0.5)$ in (2.4), and half-delayed return-to-zero (HRZ) $(\alpha, \beta) = (0.5, 1)$. The three types of DAC mentioned so far are all depicted in Figure 2.5 with their describing equations. All are easy to fabricate, for example, in an ECL-style latched comparator, a typical circuit diagram of which is shown in Figure 2.6. By diode-connecting the final differential pair rather than cross-coupling them [Gao98a], an RZ rather than an NRZ waveform is output. In the multi-feedback architecture, we can use any two of NRZ, RZ, and HRZ, or for that matter any other two different pulses, but those three types are easiest to build in a practical circuit.

Let us do a worked example that shows how to set the feedback coefficients in Figure 2.4 to give $H_{BP}(z)$ in (2.15).

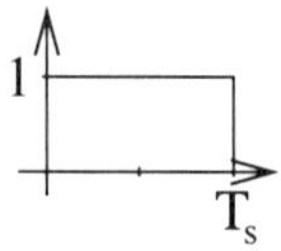

$$\hat{r}_{NRZ}(t) = \begin{cases} 1, & 0 \leq t < T_s \\ 0, & \text{otherwise} \end{cases}$$

$$\hat{R}_{NRZ}(s) = \frac{1 - e^{-sT_s}}{s}$$

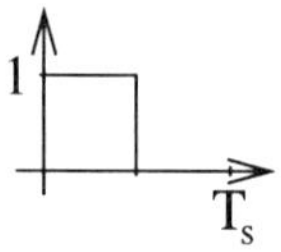

$$\hat{r}_{RZ}(t) = \begin{cases} 1, & 0 \leq t < T_s/2 \\ 0, & \text{otherwise} \end{cases}$$

$$\hat{R}_{RZ}(s) = \frac{1 - e^{-sT_s/2}}{s}$$

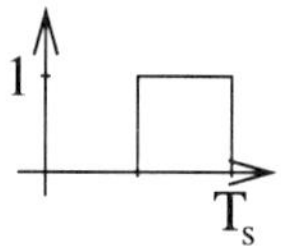

$$\hat{r}_{HRZ}(t) = \begin{cases} 1, & T_s/2 \leq t < T_s \\ 0, & \text{otherwise} \end{cases}$$

$$\hat{R}_{HRZ}(s) = e^{-sT_s/2}\frac{1 - e^{-sT_s/2}}{s}$$

Figure 2.5. Common DAC pulse types.

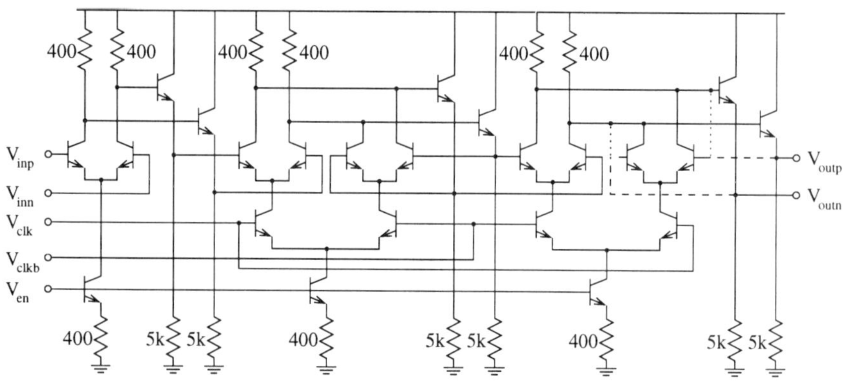

Figure 2.6. ECL-style latched comparator with preamplification for enhanced resolution at high speed. For an NRZ comparator, connect the final differential pair via the dashed lines; for RZ, connect the dotted lines instead.

Example 2.5: We choose $\omega = A = \pi/2$ in the resonator blocks. The basic idea is to use superposition on the four feedback paths. Begin with k_{2r}: the transfer function "seen" from output bit to quantizer input is simply

$$\hat{H}_{2r}(s) = k_{2r}\frac{\frac{\pi}{2}s}{s^2 + (\frac{\pi}{2})^2}. \tag{2.17}$$

Using Matlab's `c2d` function gives an equivalent DT transfer function of

$$H_{2r}(z) = k_{2r}\frac{(1-\frac{1}{\sqrt{2}})z - \frac{1}{\sqrt{2}}}{z^2+1} \tag{2.18}$$

where, as in Example 2.3, we work in the state-space domain *and* we alter the DT B matrix for non-NRZ DAC pulses similarly to (2.14):

$$B_{(\alpha,\beta)} = \hat{A}^{-1}\exp(\hat{A}(1-\alpha)) - \exp(\hat{A}(1-\beta))\hat{B}_{(0,1)}. \tag{2.19}$$

Proceeding similarly for the k_{2h} path we find

$$H_{2h}(z) = k_{2h}\frac{\frac{1}{\sqrt{2}}z - (1-\frac{1}{\sqrt{2}})}{z^2+1}. \tag{2.20}$$

k_{4r} and k_{4h} see transfer functions that involve

$$\left(\frac{\frac{\pi}{2}s}{s^2+(\frac{\pi}{2})^2}\right)^2 \tag{2.21}$$

because of the series connection of the two resonator blocks. Converting these using `c2d` gives

$$H_{4r}(z) = k_{4r}\frac{0.5077z^3 - 0.8330z^2 + 0.0476z + 0.2777}{(z^2+1)^2} \tag{2.22}$$

$$H_{4h}(z) = k_{4h}\frac{0.2777z^3 + 0.0476z^2 - 0.8330z + 0.5077}{(z^2+1)^2}. \tag{2.23}$$

We need to pick the ks so that the linear combination of the DT equivalents yield the overall desired $H_{BP}(z)$ in (2.15):

$$H_{4r}(z) + H_{4h}(z) + [H_{2r}(z) + H_{2h}(z)] \times \frac{z^2+1}{z^2+1} = \frac{2z^2+1}{(z^2+1)^2}. \tag{2.24}$$

Equating powers of z in the numerators of both sides of (2.24) and solving for the ks yields

$$\{k_{r4}, k_{r2}, k_{h4}, k_{h2}\} = \{-1.0868, -2.1339, 0.4502, 1.4874\}. \tag{2.25}$$

The Matlab code to solve this problem is listed in Figure 2.7. □

2. IMPLICIT ANTIALIAS FILTERS

In any DT system, Nyquist taught us that when the sampling rate is f_s, any two tones which differ in frequency by a multiple of f_s are indistinguishable

```
% Matlab for multifeedback modulators
  p2 = pi/2;
% Do 2nd-order NRZ system
  [a2,b2,c2,d2] = tf2ss([0 p2 0],[1 0 p2*p2]);
  sysc2 = ss(a2,b2,c2,d2);
  sysd2n = c2d(sysc2,1);
% Find RZ system
  sysd2r = sysd2n;
  sysd2r.b = inv(sysc2.a)*(expm(sysc2.a)-expm(sysc2.a*0.5))*sysc2.b;
% Find HRZ system
  sysd2h = sysd2n;
  sysd2h.b = inv(sysc2.a)*(expm(sysc2.a*0.5)-eye(size(sysc2.a)))*sysc2.b;
% Below:  (z^2+1)/(z^2+1) for multiplying 2nd order coeffs by
  tf2 = tf([1 0 1],[1 0 1],1);
  tfd2n = tf(sysd2n)*tf2;
  tfd2r = tf(sysd2r)*tf2;
  tfd2h = tf(sysd2h)*tf2;
% Do 4th-order NRZ system
  [a4,b4,c4,d4] = tf2ss([0 0 p2^2 0 0],[1 0 2*p2*p2 0 p2^4]);
  sysc4 = ss(a4,b4,c4,d4);
  sysd4n = c2d(sysc4,1);
```

Figure 2.7. Matlab code for multi-feedback $f_s/4$ BP modulator.

```
% Find RZ system
  sysd4r = sysd4n;
  sysd4r.b = inv(sysc4.a)*(expm(sysc4.a)-expm(sysc4.a*0.5))*sysc4.b;
% Find HRZ system
  sysd4h = sysd4n;
  sysd4h.b = inv(sysc4.a)*(expm(sysc4.a*0.5)-eye(size(sysc4.a)))*sysc4.b;
% Find 4th order transfer functions
  tfd4n = tf(sysd4n);
  tfd4r = tf(sysd4r);
  tfd4h = tf(sysd4h);
% Create LHS matrix to solve--assume RZ and HRZ DACs
  amat = zeros(4,4);
  num = tfd4r.num{:}; amat(:,1)=num(2:5)';
  num = tfd2r.num{:}; amat(:,2)=num(2:5)';
  num = tfd4h.num{:}; amat(:,3)=num(2:5)';
  num = tfd2h.num{:}; amat(:,4)=num(2:5)';
% Solve for feedback k's to satisfy 2*z^2+1
  kd = amat \ [0;2;0;1];
  k4r = kd(1); k2r = kd(2); k4h = kd(3); k2h = kd(4);
```

Figure 2.7 (continued). Matlab code for multi-feedback $f_s/4$ BP modulator.

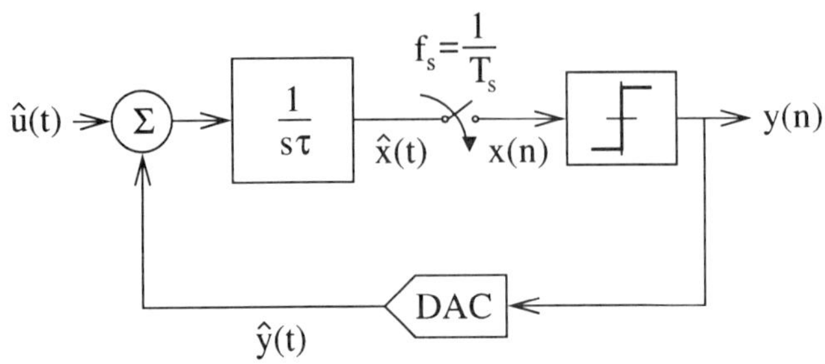

Figure 2.8. First-order CT ΔΣM.

from one another and so overlap in a spectral plot. This problem is usually referred to as signal *aliasing*. DT ΔΣMs usually require an extra filter (an *antialiasing* filter) to be placed prior to their input to bandlimit the input signal and hence reduce the problem of aliasing. A nice feature of CT ΔΣMs is, their very nature means they have some built-in antialiasing protection.

It is easiest to understand this intuitively by considering a first-order low pass CT modulator with a single integrator $1/s\tau$ and an NRZ DAC, Figure 2.8. Immediately after the sampler, the quantizer input in the z domain obeys the relation

$$x(n+1) = x(n) - \frac{T_s}{\tau} y(n) + \frac{1}{\tau} \int_{nT_s}^{(n+1)T_s} \hat{u}(t) dt. \tag{2.26}$$

Notice that the input signal is integrated over one clock period prior to being sampled. This is sometimes referred to as *boxcar integration*: the input is multiplied by a rectangular pulse (which has a shape like a railway boxcar) prior to integrating it. We may also write that input integral as a convolution of the input with a rectangular pulse:

$$\int_{nT_s}^{(n+1)T_s} \hat{u}(t) dt \quad = \quad \hat{u}(t) * \mathrm{rect}(nT_s, (n+1)T_s) \tag{2.27}$$

$$= \quad \hat{U}(s) \cdot \mathrm{sinc}(f/f_s), \tag{2.28}$$

where (2.28) follows from (2.27) because convolution in the time domain is the same as multiplication in the frequency domain. (2.28) tells us that the input spectrum is multiplied by the spectrum of a rectangular pulse, namely, a sinc function. This latter function has spectral nulls at frequencies $\pm a f_s$, $a \geq 1$. Conveniently, this is *precisely* where we want the nulls to be in LP modulators: for conversion of desired signals near dc, we are concerned about signals near multiples of the sampling frequency which would alias to near dc, and we see from (2.28) that signals near these frequencies are attenuated by the sinc. In a CT modulator, then, the antialiasing property arises because the sampling happens *after* the integrator (at the quantizer rather than at the circuit input as in a DT modulator).

Given this way of looking at things, we expect higher-order modulators to have more antialias protection (because they have more integrators before the sampler), and modulators with loop filter poles away from dc (i.e., band pass modulators) to have antialiasing at the frequencies of concern for them. In the general case, it has been shown that the implicit antialias filter for a CT $\Delta\Sigma$M can be plotted against frequency ω by evaluating [Sho96]

$$\frac{\hat{H}(j\omega)}{H(\exp(j\omega T_s))}, \tag{2.29}$$

where $\hat{H}$ is the CT modulator loop filter and H is its DT equivalent.

Example 2.6: Figure 2.9a plots the antialias filters for three example modulators: the first-order modulator in Figure 2.8, the double integration NRZ modulator, and a fourth-order $f_s/8$ band pass modulator with transfer function

$$H(z) = \frac{0.6527z^3 - 1.6923z^2 + 1.5230z - 0.6154}{(z^2 - \sqrt{2}z + 1)^2}. \tag{2.30}$$

A detail of those same plots on a logarithmic vertical scale appears in Figure 2.9b, and we verify our intuition that the second-order LP modulator has more attenuation near multiples of f_s than the first-order LP modulator. Moreover, in the $f_s/8$ modulator, the frequencies that would alias in-band are $7f_s/8$, $9f_s/8$, $15f_s/8$, etc. As predicted, these filters are nulled by that modulator's built-in antialias filter. □

Example 2.7: The discussion so far has implicitly assumed NRZ DAC pulses, but there is still antialiasing in modulators with other kinds of DAC pulse. An experimental demonstration of this fact is provided here. In [Sho96], an actual second-order multi-feedback $f_s/4$ BP modulator was characterized by injecting a 0dB input tone at the frequency listed in the first column of Table 2.2. The frequency to which this tone aliases at the clock rate of f_s = 200MHz is listed in the second column, and at that frequency, the suppression of the alias was measured. For an $f_s/4$ modulator, ideally aliases at exactly $3f_s/4$ = 150MHz and $5f_s/4$ = 250MHz would be completely attenuated; in practice, the attenuation is more than 40dB, though not infinite because of parasitic coupling, etc. Attenuation is seen to worsen as we move further away from these frequencies, as demonstrated in the previous example. □

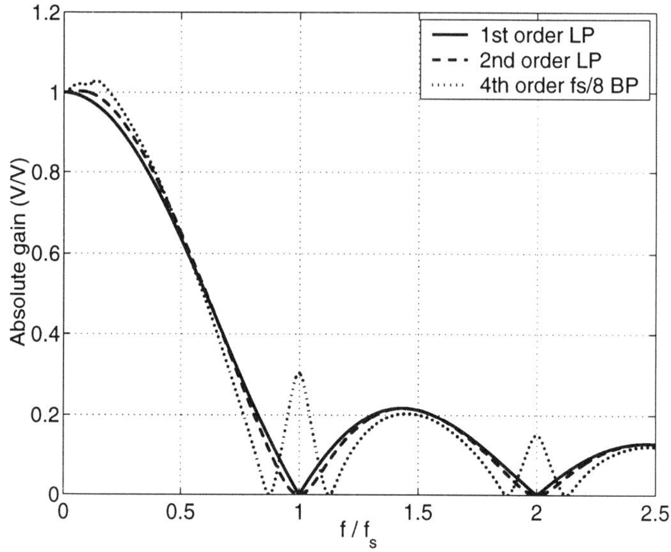

Figure 2.9a. Implicit antialias filters for various modulators.

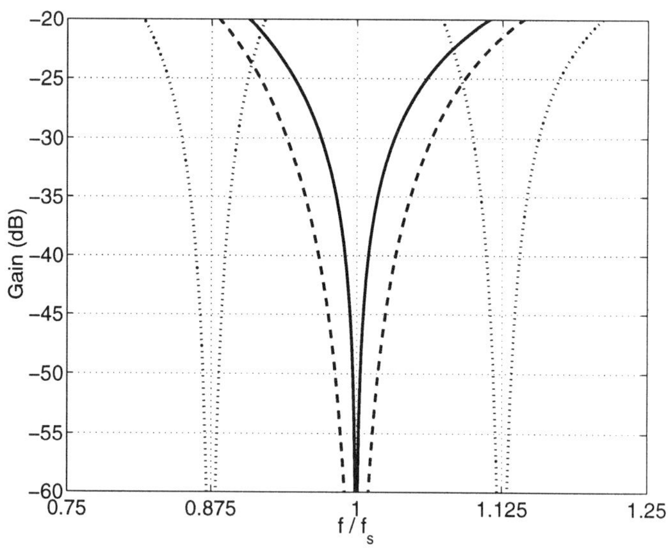

Figure 2.9b. Detail of implicit antialias filter plot.

Table 2.2. Experimental antialiasing demonstration.

Input frequency (MHz)	*Aliased frequency (MHz)*	*In-band attenuation (dB)*
149.6	50.4	34
149.8	50.2	39
150.0	50.0	42
150.2	49.8	39
150.4	49.6	34
249.6	49.6	39
250.0	50.0	49
250.4	50.4	45

3. UNEQUAL DAC PULSE RISE/FALL TIME

In general, CT $\Delta\Sigma$Ms are sensitive to the exact shape of the DAC pulse: any nonuniformities will tend to degrade performance. We will study this in the context of high-speed modulators in Chapter 5; here, we consider the specific problem where the rise time of a DAC pulse is different from its fall time, something first brought to light by Adams [Ada86].

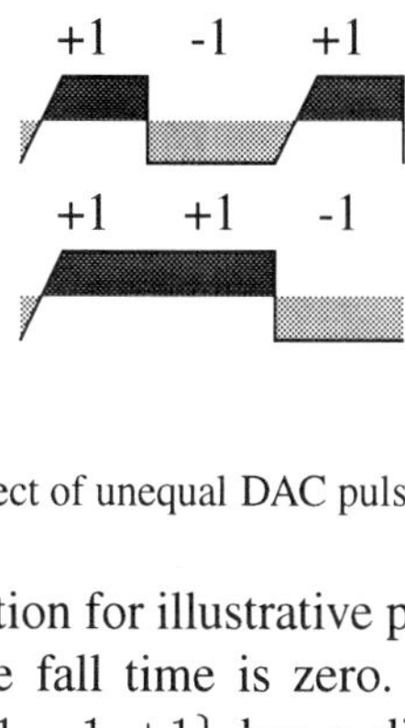

Figure 2.10. Effect of unequal DAC pulse rise and fall times.

Consider an extreme situation for illustrative purposes where the DAC pulse rise time is nonzero and the fall time is zero. In Figure 2.10, a sequence of output bits that goes $\{+1, -1, +1\}$ has a different area than a sequence $\{+1, +1, -1\}$: compare the area of the dark regions to that of the light regions. The area under the pulses is significant because the pulses are integrated inside the forward modulator loop; ideally, the ordering of the same set of bits should not change this area, but when the pulse rise and fall times differ, the area *does* change. DAC pulse errors are fed back all the way to the modulator input (i.e., they are input-referred), and so are not noise-shaped by the action of the loop.

Example 2.8: A case which can be treated analytically that demonstrates the effect of rise/fall time asymmetry is the first-order LP modulator with a dc input. In this situation, the output bit sequence is deterministic

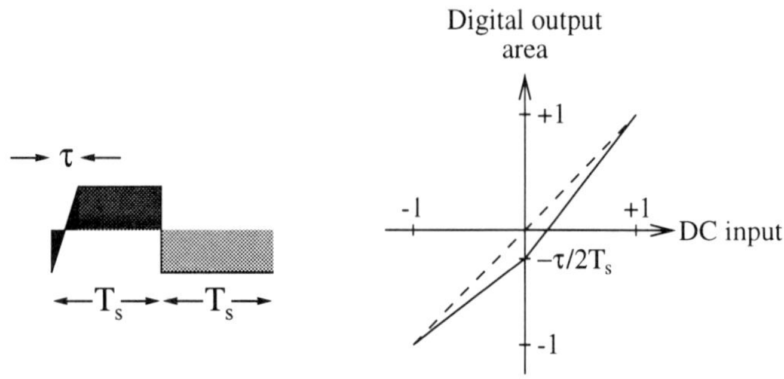

Figure 2.11. First-order modulator asymmetric DAC pulse: effect for dc input of 0V (left) and overall transfer characteristic (right).

[Fee91]; for a midrail dc input, the output alternates between $+1$ and -1. If the DAC pulses have period T_s and an asymmetry τ, the area can be found from Figure 2.11: the two black trapezoids are of equal area but opposite signs, meaning they cancel each other out, and the remaining positive (dark) and negative (light) rectangles have areas $T_s - \tau$ and T_s, respectively. As a fraction of the total pulse time $2T_s$, the area of these digital pulses can be expressed as

$$\frac{T_s - \tau - T_s}{2T_s} = \frac{-\tau}{2T_s}. \tag{2.31}$$

Carrying out a similar calculation for all dc inputs and plotting digital output area yields the solid line in the graph on the right of Figure 2.11. With uniform DAC pulses, the area of the digital output would equal the dc input, but in this case, there is a kink in the curve such that it intersects $-\tau/2T_s$.

This looks suspiciously like integral nonlinearity in a Nyquist-rate ADC—and in fact, it has a very similar effect to INL. For example, if the DAC pulses had an asymmetry of $\tau/T_s = 0.1\%$, a full-scale sine wave modulator would produce second-order harmonic distortion with magnitude -72dB. This implies the maximum modulator resolution is limited to 12 bits. Therefore, asymmetric DAC pulses limit the maximum achievable performance of a $\Delta\Sigma$M. □

Higher-order modulators produce nondeterministic output bit streams, so to study the performance effect we must resort to probabilistic means [Ada86]. In that paper, Adams showed that asymmetric DAC pulses result in white noise in the output spectrum at a level higher than that of the quantization noise. The final result of his analysis says that for a given sampling rate, OSR, and desired

SNR, the maximum asymmetry must approximately satisfy the relation

$$\tau \leq \frac{4 \cdot T_s \cdot \sqrt{\text{OSR}}}{\text{SNR}_{\text{desired}}}. \tag{2.32}$$

As an example, his part used $T_s = 163\mu$s, OSR $= 150$, and for a 16-bit audio converter (SNR $=$ 96dB), (2.32) gives $\tau < 130$ps. This was a serious challenge in 1986 with a modulator built out of discrete components!

Since that time, two solutions to the problem of asymmetric DAC pulses have been proposed. The first is to use differential rather than single-ended circuitry; it is demonstrated in [Jen95] that the sum of two asymmetric single-ended DAC waveforms produce a symmetrical differential DAC waveform. The second is to use RZ DACs rather than NRZ; then, as shown in Figure 2.12, the same area

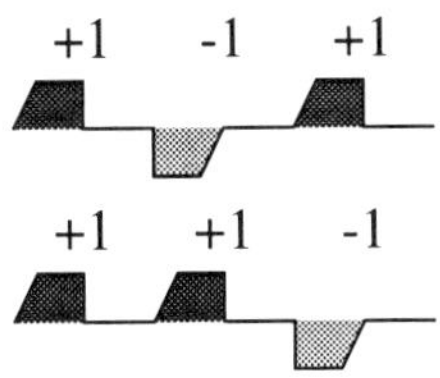

Figure 2.12. RZ DAC pulses for circumventing asymmetry problems.

is obtained for the same set of output bits irrespective of their order.

4. PRACTICAL DESIGN AND CHARACTERIZATION

The ideal modulator design procedure laid out in the first section of this chapter assumed rectangular DAC pulses. The previous section treats a slightly more practical case of trapezoidal DAC pulses. In a real circuit, the DAC pulses will exhibit overshoot and ringing. Is the ideal design procedure still useful in this case? And how can we be sure a practical CT modulator implements the $H(z)$ that gives us the noise shaping we desire?

To answer the first question, the ideal design procedure *is* still useful: carrying it out gives us at least a rough idea of how to set the feedback DAC currents. As for the second question, in §4.1 we present a procedure for use in SPICE or any other circuit simulator that helps us get the DAC feedback levels right for a given $H(z)$ and whatever actual DAC pulse shape we have. After that, §4.2 explains the dual of this procedure: given an actual CT modulator, we can find its $H(z)$ from a relatively short simulation.

4.1 SPICE-BASED FEEDBACK SETTING

In essence, we turn the modulator into the open-loop system of Figure 2.1a so that we can apply an impulse to the loop and use superposition. We assume the feedback currents are tunable, and will use the modulator from [Jen95], Figure 2.13a, to illustrate the steps of the procedure.

1. A time-domain expression

$$h(n) = \mathcal{Z}^{-1}\{H(z)\} \tag{2.33}$$

   for the desired equivalent DT impulse response is determined using the inverse $\mathcal{Z}$-transform.

2. The modulator input is grounded, and the feedback loop is broken at the quantizer input, Figure 2.13b. We shall apply our own voltage to the quantizer input in the next step; however, in the figure, a second copy of the quantizer still remains connected to the output of the final integrator (or resonator, in the case of BP modulators). This is because in a real circuit it will be necessary to keep the loading on the final integrator the same as in the real modulator so that this integrator functions as expected.

3. For each feedback current k_i:

   (a) k_i is set to a known value κ_i and all other k_is are set to zero, Figure 2.13c.

   (b) A transient simulation of the circuit is executed for m clock cycles, where m is the modulator order. The quantizer which drives the DACs is clocked and an input voltage applied to it such that it produces a single impulse, followed by a string of zeros, Figure 2.13c.

   (c) The output of the final integrator block is measured at sampling instants, Figure 2.13d, and the voltage values $x_i(T_s), x_i(2T_s), \ldots, x_i(mT_s)$ are noted.

4. The system of m equations

$$k_1 \frac{x_1(iT_s)}{\kappa_1} + \cdots + k_m \frac{x_m(iT_s)}{\kappa_m} = h(i), \; i = 1, \ldots, m \tag{2.34}$$

   is solved for the m unknown DAC currents $\{k_1, \ldots, k_m\}$.

A little thought will convince the reader why this works. To crystallize the point, we use a worked numerical example.

Example 2.9: Suppose we are designing a double integration modulator. Then we find

$$H(z) = \frac{2z - 1}{(z-1)^2} \quad \rightarrow \quad h(n) = n + 1 - \delta(n) \tag{2.35}$$

where H(z) is from (2.7) and $\delta(n)$ is the DT unit impulse function. Let us take the architecture to be that in Figure 1.15 from [Jen95], and assume parameters of $f_s = 1$GHz, $g_m = 1$mA/V, and $C = 2$pF.

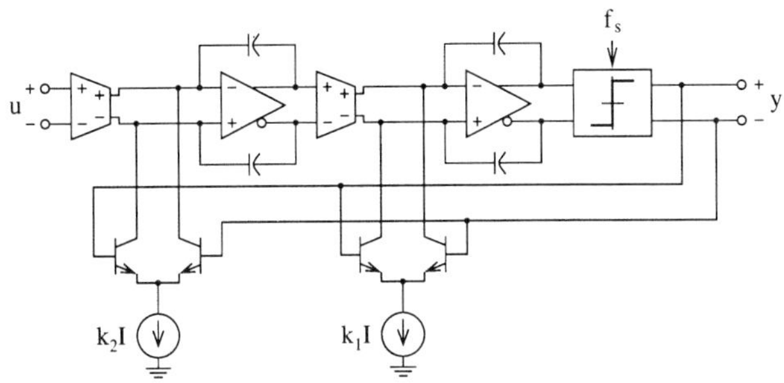

Figure 2.13a. Starting position for SPICE-based feedback current setting.

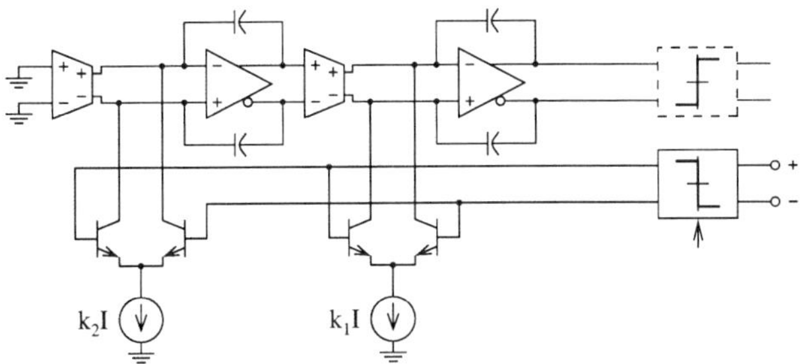

Figure 2.13b. Zeroed input, quantizer disconnected from loop and dummy quantizer connected for proper loading.

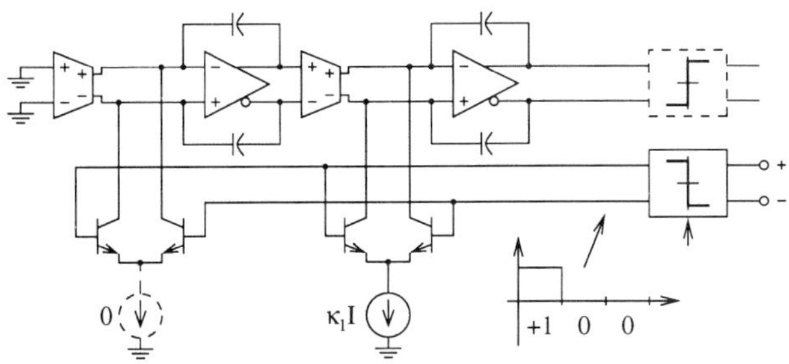

Figure 2.13c. One DAC enabled, all others disabled, quantizer clocked to send impulse into loop.

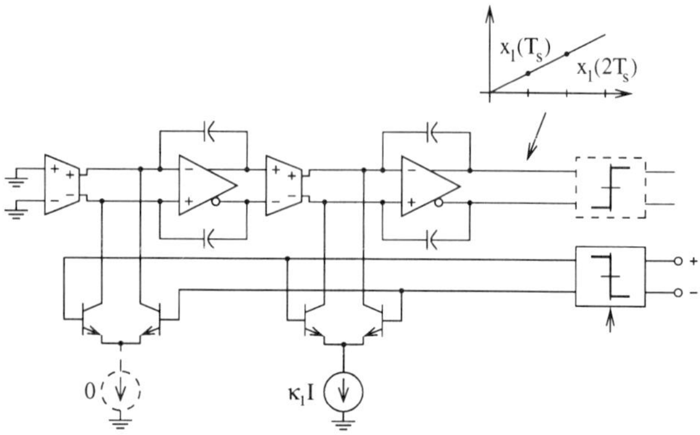

Figure 2.13d. Quantizer input voltage measured at sampling instants.

If we implement the circuit using ideal models in SPICE and RZ feedback, then follow the procedure by zeroing the input and pulsing the quantizer with one DAC enabled at a time, we find:

- With $k_1 = \kappa_1 = 0.5\text{mA}$ and $k_2 = 0$, we measure $x_1(T_s) = 125\text{mV}$ and $x_1(2T_s) = 125\text{mV}$.
- With $k_2 = \kappa_1 = 0.4\text{mA}$ and $k_1 = 0$, we measure $x_2(T_s) = 37.5\text{mV}$ and $x_2(2T_s) = 87.5\text{mV}$.

This leads us to a system of equations

$$h(1) = 2 = k_1 \frac{125}{0.5} + k_2 \frac{37.5}{0.4} \qquad (2.36)$$

$$h(2) = 3 = k_1 \frac{125}{0.5} + k_2 \frac{87.5}{0.4}. \qquad (2.37)$$

Solving (2.36) and (2.37) yields

$$k_2 = 8\text{mA}, \; k_1 = 5\text{mA}. \qquad (2.38)$$

The $h(n)$ values in (2.36) and (2.37) are in volts; 2V and 3V might be too large in a nonideal circuit, so we could scale down k_1 and k_2 by a factor of, say, 10 and achieve a factor of 10 reduction in $h(n)$ to 0.2V and 0.3V. □

4.2 SPICE DESIGN DUAL: *Z*-DOMAIN EXTRACTION

If it is possible in SPICE to determine how to set the DAC currents to get a particular $H(z)$, then the opposite procedure should be possible: for a given DAC current setting, what $H(z)$ does a modulator implement? During the course of our work with CT $\Delta\Sigma$Ms, we have been asked exactly this question for a practical high-speed double integration modulator [Sch96b]; in response, we developed z-domain extraction [Che98a]. Before we explain the procedure, we provide some background.

INITIAL OBSERVATIONS

An early prototype of the modulator in question was very similar in design to the modulator in Figure 1.15. It clocked at $f_s = 1$GHz, and a 16000-point Hann-windowed spectrum from a transistor-level SPICE simulation is shown in Figure 2.14a. The input tone was -4dB, and the SNR achieved for an OSR of 32 was 44.2dB. We show in (3.3) that an ideal double integration $\Delta\Sigma$M achieves $\text{SNR}_{\text{max}} = 56$dB, and this happens at an input level of around -4dB. Therefore, the prototype appeared nonideal by about two bits. Moreover, at an OSR of 64, the SNR only improved 3dB to 47.3dB—this shows that the baseband noise is white rather than shaped at 15dB per octave of oversampling, as we would expect for a second-order $\Delta\Sigma$M.

A C program was written to model the SPICE circuit, including such things as finite op amp gain and input resistance, nonzero delay between quantizer clock and feedback DAC output, and nonzero DAC pulse rise time. A typical spectrum for the same input conditions appears in Figure 2.14b. The spectrum looks similar except towards dc, where it continues to descend at 15dB/oct. Both modulators had an unjittered clock, and past experience with SPICE taught us that it didn't seem reasonable to attribute the SPICE results to, e.g., roundoff error. The main question we were interested in answering was, why does the spectrum of the SPICE simulation become white? We wondered if it might be a problem with pole placement of the equivalent $H(z)$—poles far inside the unit circle would result in a flat bottom in the output spectrum.

Z-DOMAIN EXTRACTION

Our method for finding $H(z)$ exploits the bidirectionality of the DT/CT equivalence in §1.1. For the general DT modulator in Figure 1.16 on page 25, we had a time-domain expression for the quantizer input $x(n)$ in (1.12), reproduced here:

$$x(n) = \sum_{k=1}^{m} a_k x(n-k) + \sum_{k=1}^{m} b_k u(n-k) + \sum_{k=1}^{m} c_k y(n-k). \tag{2.39}$$

We are using the impulse-invariant transformation for DT/CT equivalence, which enforces the condition (2.1):

$$x(n) = \hat{x}(t)|_{t=nT_s}. \tag{2.40}$$

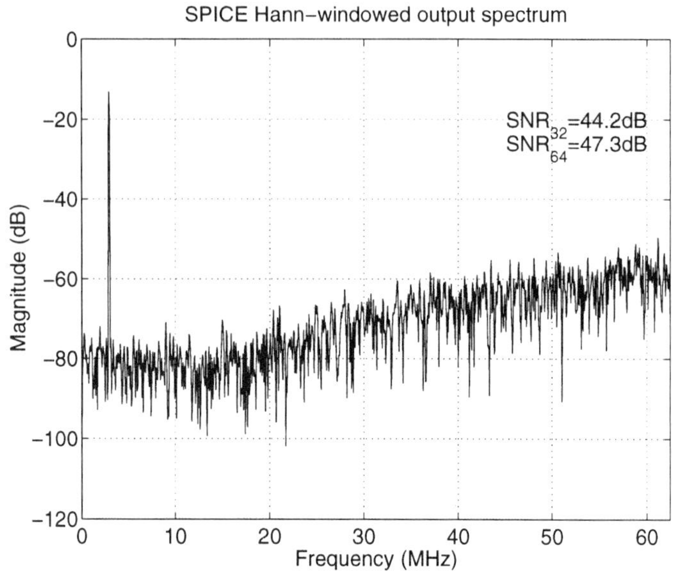

Figure 2.14a. Typical output spectrum for SPICE prototype.

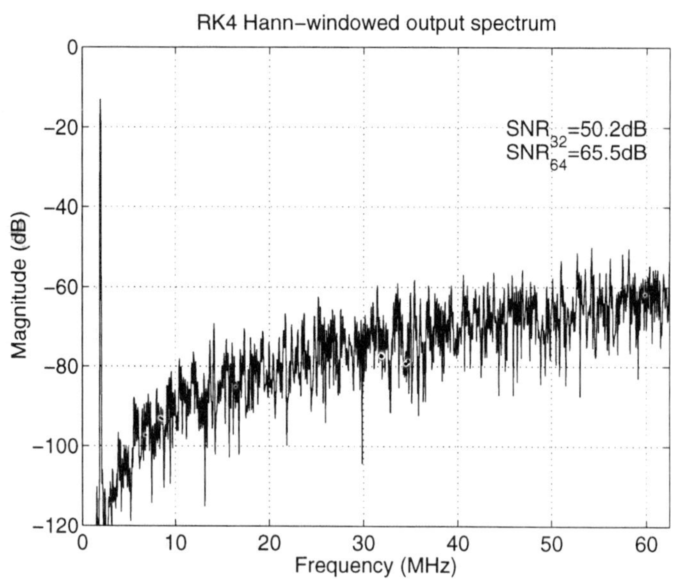

Figure 2.14b. Typical output spectrum for C program using the same parameters.

If (2.39) holds for a DT modulator, then it must also hold for a CT modulator at sampling instants:

$$\hat{x}(nT_s) = \sum_{k=1}^{m} \left[a_k \hat{x}((n-k)T_s) + b_k \hat{u}((n-k)T_s) + c_k \hat{y}((n-k)T_s)\right]. \tag{2.41}$$

This suggests the following: if we simulate a CT ΔΣM and extract the quantities $\{\hat{x}(nT_s), \hat{u}(nT_s), \hat{y}(nT_s)\}$, then we ought to be able to find $\{a_k, b_k, c_k\}$ such that (1.15) is satisfied. This will give us the DT equivalent for the CT modulator.

Example 2.10: During the 16000-clock cycle SPICE simulation of the prototype, the values of $\hat{x}$, $\hat{u}$, and $\hat{y}$ at sampling instants were printed out. Using a least-squares fitting approach in Matlab with the `tls` function, the following best-fit DT difference equation was found for the group of 50 consecutive samples of $\hat{x}(nT_s) = x(n)$ beginning at sample 700:

$$\begin{aligned} x(n) &= 1.9835x(n-1) - 0.9886x(n-2) \\ &+ 0.2319u(n-1) - 0.2083y(n-1) \\ &+ 0.0511y(n-2) + 0.0462y(n-3), \end{aligned} \tag{2.42}$$

$$\|\varepsilon\|_\infty = 19.43\text{mV}, \quad \|\varepsilon\|_2 = 4.914\text{mV}. \tag{2.43}$$

The fit is not perfect, as evidenced by the nonzero $\|\varepsilon\|_\infty$ and $\|\varepsilon\|_2$ values in (2.43), which are (respectively) the maximum and rms errors between the best-fit $x(n)$ in (2.42) and the $\hat{x}(nT_s)$ from SPICE. $x(n)$ spans a range of about ±500mV, so the rms error $\|\varepsilon\|_2$ is about 1% of the full-scale range of $x(n)$.

Taking the $\mathcal{Z}$-transform of (2.42) and using (1.11), we find the loop transfer function to be

$$H(z) = 0.1042\frac{-2.0000z^{-1} + 0.4906z^{-2} + 0.4436z^{-3}}{1 - 1.9835z^{-1} + 0.9886z^{-2}}. \tag{2.44}$$

Thus, the z-domain extraction method allows us to see the $H(z)$ actually implemented (as opposed to the $H(z)$ we thought we had implemented) in a CT ΔΣM.

(2.44) is quite a bit different from the $H(z) = (-2z^{-1} + z^{-2})/(1 - z^{-1})^2$ we desire in a double integration ΔΣM. First, the ratio of the z^{-1}/z^{-2} numerator coefficients is closer to -4 than -2, and z^{-3} is nonzero. Second, from (1.2) the poles of $H(z)$ are the zeros of NTF(z), and factoring the denominator of (2.44) gives poles $z = 0.994\angle\pm4.1°$. For an ideal double integration modulator, this would be $z = 1\angle0°$. The effect of the practical pole values is to move the quantization noise notch from dc to $4.1f_s/360$, and to lessen the noise notch depth.

As an experiment let us try using a different group of 50 samples for the best-fit equation, starting at sample 7000. This yields a best-fit $x(n)$ of

$$\begin{aligned} x(n) &= 1.9587x(n-1) - 0.9832x(n-2) \\ &+ 0.2241u(n-1) - 0.2027y(n-1) \\ &+ 0.0501y(n-2) + 0.0468y(n-3), \end{aligned} \tag{2.45}$$

$$\|\varepsilon\|_\infty = 26.62\text{mV}, \quad \|\varepsilon\|_2 = 7.061\text{mV}. \tag{2.46}$$

The NTF zeros are now found to be $0.992\angle\pm 9.0°$. The magnitude is similar to that found from (2.44), but the angle has changed from $4.1°$ to $9.0°$. The coefficients of the best-fit equation seem a little bit sensitive to the group of samples chosen. □

EXAMINING THE ERRORS

The realized pole locations are quite close to the unit circle, hence this does not seem to explain the flat bottom on the SPICE output spectrum. We thus took a closer look at the difference between the simulated and extracted $H(z)$. For the groups of samples in Example 2.10, Figure 2.15 illustrates the SPICE values of $\hat{x}(nT_s)$ with clear bars, and the errors (the solid bars, magnified for ease of viewing) between $\hat{x}(nT_s)$ and the best-fit Matlab equations $x(n)$ in (2.42) and (2.45). At samples 709 and 710 in the top graph of Figure 2.15, there is a large error followed by an error of opposite sign; the same is true at samples 7018 and 7019 in the bottom graph. A third instance of the same problem at samples 179 and 180 was found by resimulating with a smaller printing time step; the outputs of the first and second integrators are plotted in Figure 2.16.

This design is clocking sufficiently quickly that there is a small delay between the clock zero crossing and the output of the DAC changing direction. We denote this *excess loop delay*, and it is the subject of Chapter 4 of this monograph. Nominally in this design, the excess delay is about $0.2T_s$, 20% of a clock period. We noticed the matching errors of opposite sign coincided with *additional* excess delay at sample 178: in Figure 2.16, we see the excess delay is about $0.3T_s$. Recall that the second integrator output in Figure 1.15 is precisely $\hat{x}$, the quantizer input; notice that in Figure 2.16(b), the voltage at sample 178 (illustrated by the small circle) is close to zero.

It is this which causes the additional excess delay: the quantizer is a latched comparator with a finite regeneration gain, so a small quantizer input leads to a longer quantizer regeneration time. Figure 2.17 illustrates the point further by plotting excess delay against quantizer input magnitude for many sampling instants [Che97]. It is this *metastability* in the quantizer which leads to the white noise floor in the SPICE simulation. We cover this in more detail in Chapter 5.

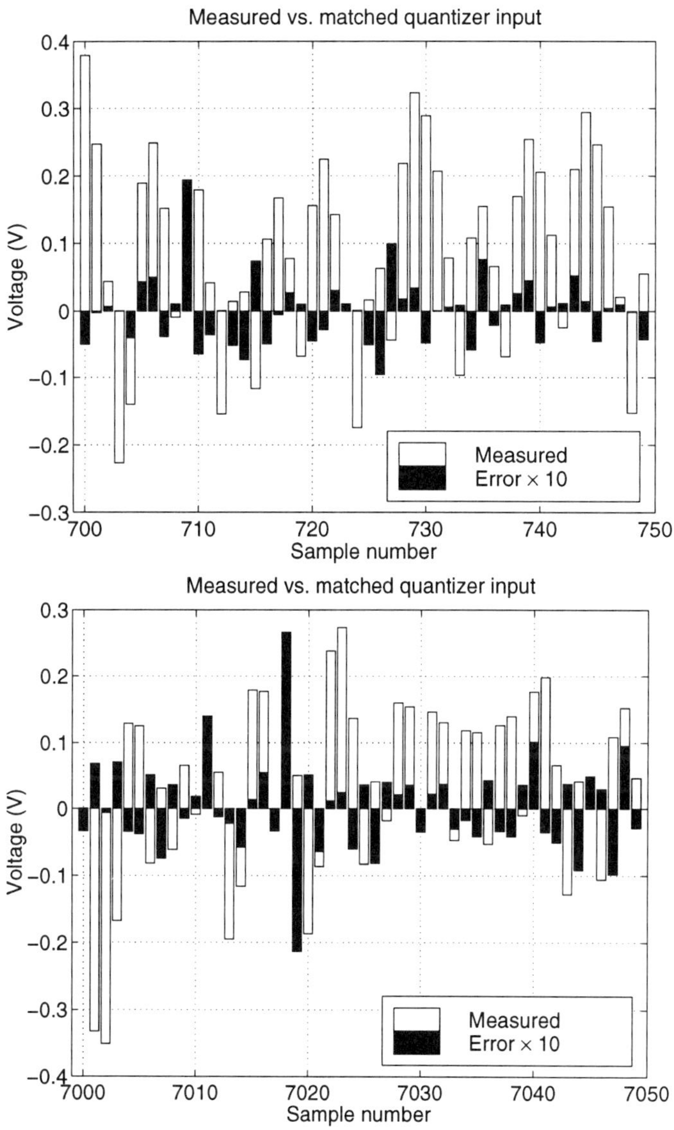

Figure 2.15. Examples of z-domain extraction from SPICE data.

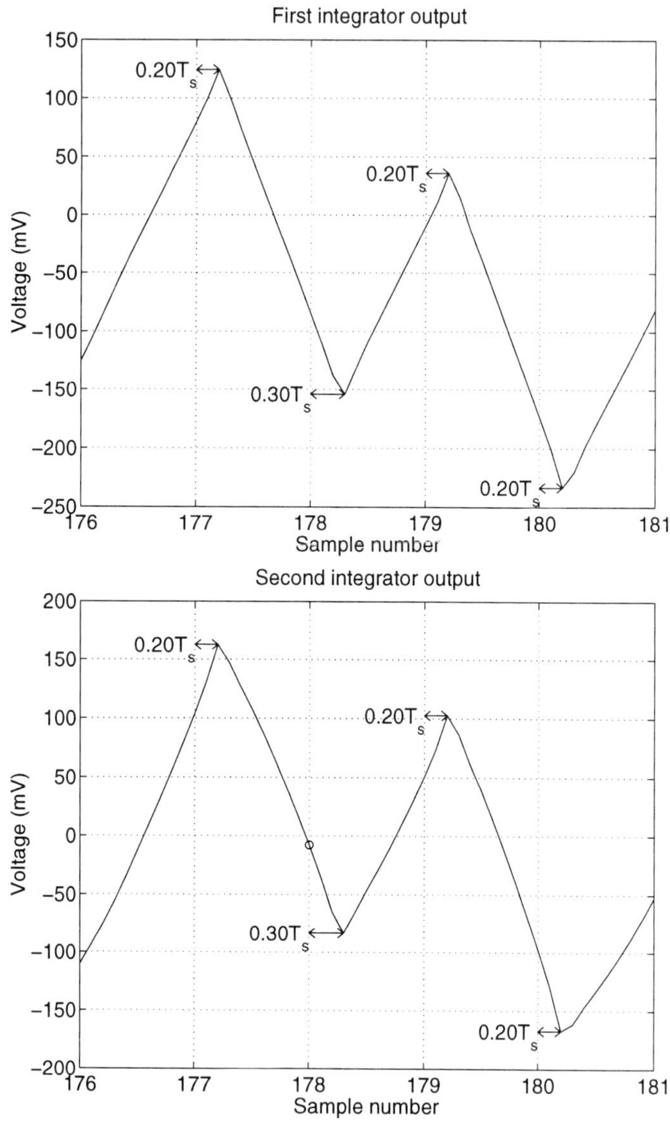

Figure 2.16. First and second integrator output waveforms from SPICE showing additional excess delay at sample 178.

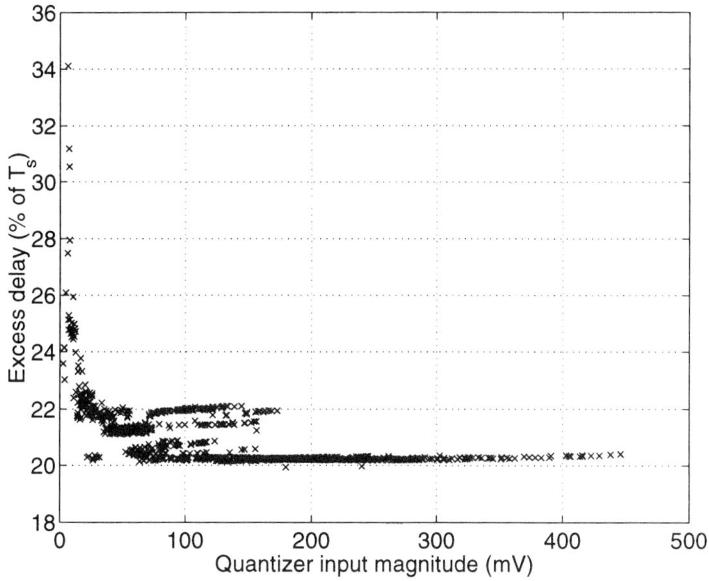

Figure 2.17. Effect of quantizer metastability on excess loop delay.

USEFULNESS OF Z-DOMAIN EXTRACTION

Because of the z-domain extraction method, we stumbled upon the significance of metastability in CT $\Delta\Sigma$Ms. In general, what are the good and bad points of the method? The good things are:

- It requires relatively few samples to work, and hence relatively little simulation: 50 samples are enough for a good least-squares fit. These 50 samples could be the first 50 rather than later sequences of 50 as we used in Example 2.10, which means we can apply it with a quick SPICE simulation rather than a lengthy one.

- It works on data from any simulation program, SPICE or otherwise, which can print out circuit values at sampling times.

- It turns out to be good at modeling certain types of nonideality, such as DAC waveforms with delay or nonzero rise time.

- It allows us to determine the $H(z)$, and consequently the NTF(z), actually implemented. This could be useful as a design check on NTF(z).

The bad things are:

- Guesswork is sometimes required as to which terms to include on the right-hand side (RHS) of the fitting equation: better guesses yield smaller matching errors.

- Certain types of nonidealities, such as nonlinear integrating capacitors (or, indeed, quantizer metastability—it was this that was predominantly to blame for the largish matching errors in (2.43) and (2.46)), seem difficult or impossible to model exactly. This might be improved with a better selection of fitting terms.

We had, perhaps naively, hoped to be able to use the method to supplant CT simulation altogether: if we could obtain a perfect z-domain fit for a CT modulator, then we could simply simulate a CT modulator using the difference equation for that $H(z)$. Instead, the method seems appropriate to use *in conjunction with*, rather than in place of, full-circuit CT simulation.

We should also mention that it can be, and was successfully, used in an SC modulator to identify clipping integrator outputs as the reason for poor performance, so it can be applied to DT simulations as well.

5. SUMMARY

The best way to choose a loop filter for a CT $\Delta\Sigma$M is to choose a DT loop filter which meets the performance specifications, and then transform it using the impulse-invariant transformation in Matlab to the equivalent CT filter based on the CT $\Delta\Sigma$M DAC pulse shape. Low pass $\Delta\Sigma$Ms are built with CT integrators $1/s$, and if we wish to build band pass $\Delta\Sigma$Ms with purely band pass CT resonators $s/(s^2 + \omega^2)$, then we should use two feedback DACs with different pulse shapes to give us full control over the noise-shaping behavior. CT modulators are nice because they sample the signal at the quantizer rather than at the input, which gives them some built-in antialiasing protection that improves with modulator order. Care must be taken with the shape of the DAC pulses: they need not be perfectly rectangular, but they should be as uniform as possible—every DAC pulse should look as much like every other DAC pulse. In a practical circuit with nonideal DAC pulses, there are simulation methods which tell us how to scale the feedback currents to achieve certain noise-shaping behavior; conversely, we can verify that our modulator builds the right equivalent DT filter using the z-domain extraction technique.

Chapter 3

DELTA–SIGMA MODULATOR IMPLEMENTATION ISSUES

The theory of ideal delta–sigma modulators is quite well-understood [Nor97, Chap. 4–5]. The purpose of this chapter is twofold: first, to summarize the papers from the published literature which discuss the effects of commonly-encountered nonidealities on the performance of $\Delta\Sigma$Ms, and second, to list the important literature papers regarding CT $\Delta\Sigma$Ms specifically, of which there are considerably fewer than those that discuss DT $\Delta\Sigma$Ms. Since this monograph is largely about high-speed CT $\Delta\Sigma$Ms, we include a summary of the performance achieved in high-speed CT $\Delta\Sigma$Ms published in the open literature to date.

1. NONIDEALITIES IN $\Delta\Sigma$MS

There are certain considerations that apply to the design of both DT *and* CT modulators. First of all, as we have seen, the problem of choosing the CT loop transfer function $\hat{H}(s)$ can be formulated in the DT domain. Additionally, there are certain nonidealities which adversely affect the performance of DT $\Delta\Sigma$Ms which have a similar effect in CT $\Delta\Sigma$Ms.

In this section, we take it as given that how to select a transfer function to achieve a given performance is understood. We survey the literature on the performance effect of nonidealities in delta–sigma modulation and summarize the results that are germane to the design of single-stage CT $\Delta\Sigma$Ms. A version of this summary for DT $\Delta\Sigma$Ms appears in [Nor97, Chap. 11], something which our summary below extends to CT $\Delta\Sigma$Ms.

1.1 OP AMPS

Not all modulators include op amps, though many do. If an op amp inside a $\Delta\Sigma$M deviates from ideal, performance is invariably worsened. Let us study the effect of various types of commonly-encountered op amp problems.

FINITE OP AMP GAIN

Probably the most widely-studied nonideal effect is that of finite op amp dc gain [Hau86, Bos88, Cha90, Fee91, Can92b, Cha92]. An ideal integrator has a DT transfer function $F(z) = 1/(z-1)$; it can be shown that an integrator built from an op amp with dc gain A_0 results in a transfer function

$$F(z) = \frac{1}{z - p(1 - 1/A_0)} \tag{3.1}$$

where p is a constant. Finite op amp gain causes *leaky integration*: the NTF zeros are moved off the unit circle towards $z = 0$, which reduces the amount of attenuation of the quantization in the baseband and therefore worsens SNR. The equivalent problem in a BP modulator occurs when the resonators have finite Q.

A good rule of thumb which applies to both DT and CT $\Delta\Sigma$Ms is that the integrators should have $A_0 \approx$ OSR, the oversampling ratio [Hau86, Bos88, Cha92, Can92b, Ber96]. If this holds, the SNR will be only about 1dB worse than if the integrators had infinite dc gain [Bos88]. In [Jen95], which is a CT $\Delta\Sigma$M using the circuit in Figure 1.15, it was shown that the parameter which limited the baseband noise floor was $A_0 R_{in} C$, where A_0 and R_{in} are the gain and input impedance of the op amp and C is the integrating capacitor. That is, the baseband noise went from shaped to white at a frequency given by $f = (2\pi A_0 R_{in} C)^{-1}$. Therefore, in CT $\Delta\Sigma$Ms, high A_0 is beneficial just as it is in DT $\Delta\Sigma$Ms.

FINITE BANDWIDTH (NONZERO SETTLING TIME)

Usually, it is assumed that an op amp can be modeled as a single-pole system with time constant τ [Hau86, Bos88, Cha90, Med94]. [Bos88] notes that for many sampled-data analog filters, the unity-gain bandwidth of an op amp must be at least an order of magnitude higher than the sampling rate; [Gre86] says it should be five times higher.

This requirement is greatly relaxed in DT $\Delta\Sigma$Ms. Both [Med94] and [Bai96] contend that incomplete integrator settling is the same as a gain error, which results in increased baseband quantization noise. However, [Hau86] finds that even with a settling error as large as 10%, 14-bit performance can be achieved as long as the settling error is *linear* (that is, exhibits an $\exp(-t/\tau)$-type behavior). The fabricated design in [Bos88] exhibited negligible performance loss for $\tau \leq T_s/2$. It thus seems that τ can be on the order of T_s for acceptable performance.

Chan [Cha92] found something similar for a CT $\Delta\Sigma$M: op amp bandwidths could be as low as f_s, the sampling frequency, and still give negligible performance loss.

FINITE SLEW RATE

Generally, in DT circuits we are worried about slewing of the *input* signal. A DT $\Delta\Sigma$M, however, is oversampled, which means the input signal slews slowly compared to the sampling rate. Thus, what concerns us is slewing of *internal* signals (most particularly op amp outputs).

It might appear that slew-rate limiting of these signals should not make any difference on top of that made by imperfect settling—so long as the outputs are "close enough" to the correct values after a full clock period, why does it matter whether they approach these values by slewing rather than linear settling? In fact, it *does* matter because op amp slewing is a *nonlinear* settling process [Cha90], and this introduces input signal harmonics in the output spectrum which degrades SNDR [Med94]. In [Bos88], a large increase both in quantization noise and harmonic distortion was observed when the slew rate dropped below $1.1\Delta/T_s$, where Δ is the difference between adjacent quantizer output levels. Note well, however, that this is an *extremely* relaxed requirement compared with non-oversampled circuits—slew rate is one of many parameters in which $\Delta\Sigma$Ms are quite tolerant of imperfections.

We show by example that a similar thing happens in CT $\Delta\Sigma$Ms.

> **Example 3.1**: Typical integrator and quantizer output waveforms for an ideal CT double integration modulator with a small dc input are depicted in Figure 3.1. The first integrator operates on the sum of two currents, a constant one determined by the quantizer bit and a nearly-constant one from the input (which in general is slowly-varying compared to the sampling clock). Hence, the integrator output $\hat{x}_1(t)$ appears as a straight line. The output $\hat{x}_2(t)$ is the integral of the sum of $\hat{x}_1(t)$ (a straight line) and the output bit (a constant), so it has a slightly parabolic shape.
>
> A typical output spectrum for the ideal modulator appears as in Figure 3.2a. For the modulator parameters chosen, the maximum slew rate required for the $f_s = 1$GHz sampling clock is about 0.35V/ns. If we limit the slew rate to 0.25V/ns, the graph in Figure 3.2b results. We see a large increase in harmonic distortion, as well as a slight increase in baseband quantization noise. □

Clearly, avoiding slew-rate limiting in CT $\Delta\Sigma$Ms is as important as in DT circuits, though doing so is not usually difficult.

LIMITED OUTPUT SWING

An mth-order $\Delta\Sigma$M has m states whose values at sampling instants completely determine modulator behavior. It is usually the case that the integrator output voltages are precisely the system states. Therefore, if the integrators are built with op amps whose output swing is not large enough to produce the required state values, modulator behavior will be altered.

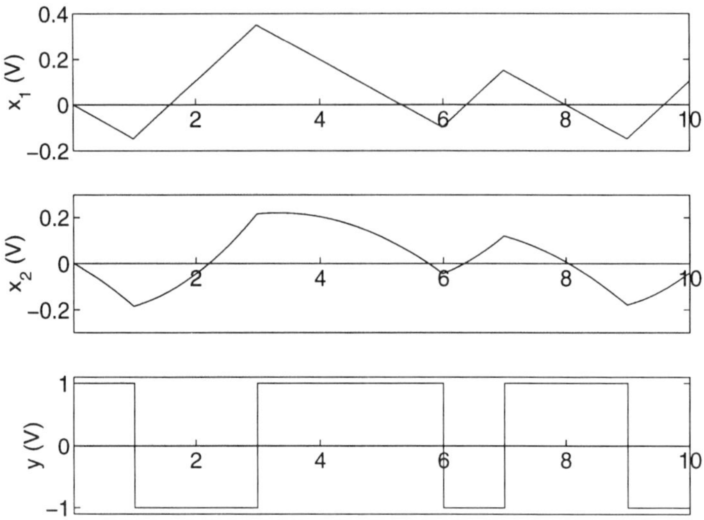

Figure 3.1. Typical integrator output and quantizer waveforms.

Both [Hau86] and [Bos88] illustrate that clipping the integrators results in severe baseband noise penalties. Fortunately, this problem has been very well-studied, and it is not difficult to scale the parameters in a ΔΣM to avoid clipping op amp outputs (e.g., [Cha90] among others). Circuit noise considerations yield a practical lower limit on how small signal swings can be. We take up this issue further in Chapter 6.

GAIN NONLINEARITY

If the gain of the op amp depends in a nonlinear manner on the op amp input voltage, harmonic distortion of large input signals appears in the output spectrum [Bos88, Med94, Dia94, Ber96]. It is difficult to give general results for how much nonlinearity can be tolerated. We *can* state that op amp gain should be made as independent of input signal level as possible, though the amount of independence required depends on the desired modulator resolution. Gain nonlinearity in the op amp nearest the input stage has the greatest effect because in both DT and CT ΔΣMs, later-stage gain nonlinearities are divided by the (large) gains of earlier stages when referred to the input [Bos88].

1.2 MISMATCH AND TOLERANCE

A traditional notion about ΔΣMs (as compared with Nyquist-rate ADCs) is that their components need not match better than the desired resolution of the

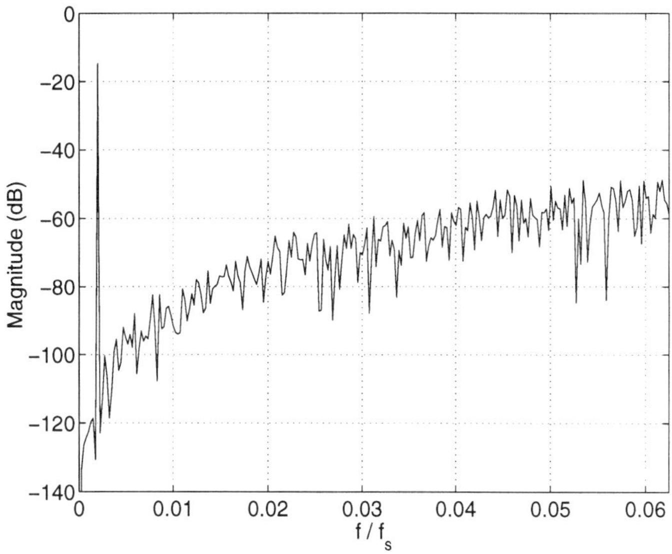

Figure 3.2a. Output spectrum using op amps with no slew-rate limiting.

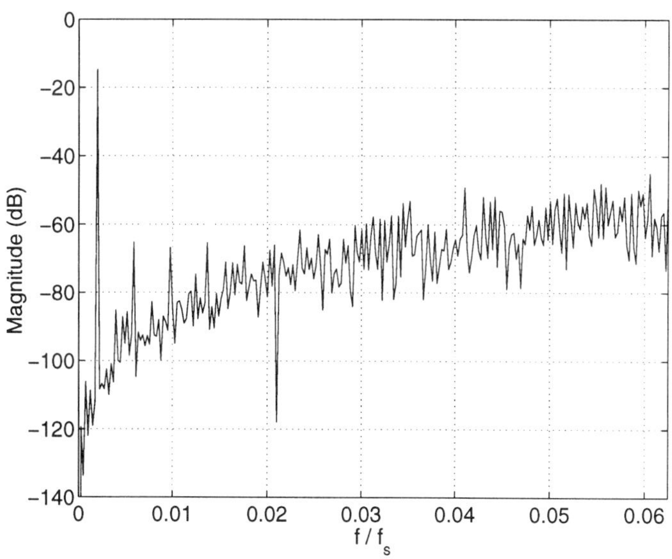

Figure 3.2b. Output spectrum using op amps with slew-rate limiting.

converter. This is true for mismatch among components in the forward loop (i.e., in the loop filter), but not true for DAC level mismatch.

COMPONENT MISMATCH AND TOLERANCE

In SC ΔΣMs, a mismatch between sampling and integrating capacitors in an SC integrator stage results in a gain error [Reb89, Baz96] whose effects we can treat in a manner similar to [Bos88]. There exist layout techniques to keep integrated capacitors matched quite well [Reb89]; as well, using large capacitors and/or clever circuit architectures can alleviate problems [Baz96].

To give an idea of the required tolerances, in a particular 90dB SC converter, it was found 5% error in individual coefficients of the loop filter led to performance losses of only 1–3dB [Cha90]. Sensitivity to tolerance obviously depends on the exact circuit architecture, so it is difficult to generalize. However, for a typical CT ΔΣM, matching requirements are unlikely to be terribly stringent, just as was found in [Cha90].

MULTIBIT DAC LEVEL MISMATCH

ΔΣMs frequently employ a one-bit quantizer for two reasons: it is easy to build, and because a feedback DAC with only two levels is inherently linear [Sch93]. If we choose to build our modulator with a multibit quantizer, then we require a multibit DAC, and now any errors in the spacing between DAC levels are directly input-referred. Thus, it would appear the resolution of the overall modulator depends directly on the DAC matching.

Fortunately, there exist techniques called *dynamic element matching* where mismatched DAC elements are "shuffled" so that different elements are used each time the same output code occurs. A survey of this area alone is quite interesting, though for brevity and relevance reasons we omit it. The most important papers which discuss DEM techniques are [Car87, Bai95, Kwa96, Jen98, Shu98], and [Nor97, Chap. 8] contains a good summary of current knowledge in the area. This same chapter also goes into digital-post correction schemes which can also compensate for multibit DAC errors.

All DAC error correction schemes could in theory be applied just as well to CT designs, something we discuss more in Chapter 4 §5..

1.3 QUANTIZERS

Just as ΔΣMs are tolerant of mismatch, so too are they tolerant of common quantizer imperfections [Hau90]. The quantizer is preceded by several high-gain stages, so dc offsets (or level spacing errors in a multibit quantizer) are negligible once input-referred. Of course, the comparator must be "fast enough" to resolve its input signal to the desired logic level; there is some discussion on this point for DT designs in [Hau90], and for CT designs not much has been said. We devote considerable attention to this important point in Chapter 5.

It transpires that for one-bit quantizers, hysteresis is not terribly problematic. Boser [Bos88] shows that hysteresis may be as severe as 0.1Δ (one tenth of the step size) with negligible performance loss in his SC circuit, though Chan [Cha92] found a requirement of 0.01Δ in his CT circuit. We consider hysteresis briefly in Chapter 5, and we discover that CT $\Delta\Sigma$Ms are very tolerant of it.

1.4 CIRCUIT NOISE

In simulation, the in-band noise floor in a $\Delta\Sigma$M output spectrum is determined only by quantization noise in an ideal modulator. Certain nonidealities like DAC level mismatch (discussed above) and clock jitter (discussed below) can also contribute to in-band noise in a simulation. In manufactured circuits, often it is the input-referred electronic circuit noise that limits performance [Bos88].

Once again, circuit noise depends on the circuit architecture. In a typical SC $\Delta\Sigma$M, noise comes from three main sources [Dia92a, Dia92b].

1. Switch resistance means the voltage sampled onto the input capacitor has uncertainty kT/C [Gre86] where k is Boltzmann's constant, T is absolute temperature, and C is the capacitance. Depending on resolution, this might require relatively large input capacitors—for particularly high-resolution converters, integrating such capacitors onto a chip might be problematic [dS90]. Off-chip capacitors could be used, or using a CT integrator as the first stage with DT integrators for later stages also works [dS90].

2. The thermal noise of the first op-amp must be kept small. It is inversely proportional to the transconductance g_m of the input MOS differential pair, which can be controlled by sizing the input devices appropriately [Gre86].

3. MOS transistors also have $1/f$ noise [Gre86], where low-frequency noise increases as 10dB/dec with decreasing frequency. This can be overcome with so-called *chopper stabilization* [Gre86, dS90] where the $1/f$ noise is cleverly modulated to the sampling frequency and thus filtered out by the decimator. As well, it is unlikely that $1/f$ noise would affect a band pass $\Delta\Sigma$M, since then the baseband would be away from low frequencies.

We leave a discussion of thermal noise in CT $\Delta\Sigma$Ms for Chapters 6 and 7.

1.5 OTHER NONIDEALITIES

There are a few other nonidealities which have been studied in connection with $\Delta\Sigma$Ms. Two effects which matter in SC designs, but not in CT designs, are nonzero switch "on" resistance and signal-dependent charge injection. The first of these limits the maximum modulator clock rate because of the RC

time constant involving the switches and the sampling capacitor [Hau90], hence small-resistance switches are often important. The second disturbs the voltage on the sampling capacitors, though it can be circumvented with techniques such as differential circuitry [Bos88], additional clock phases [Bos88, Baz96], and bottom-plate sampling [Baz96].

DESIGNING THE FIRST STAGE

We have already alluded to this in the previous subsections, but we say it explicitly here: the first stage is the most important to design well in terms of its thermal noise, linearity, matching, etc. This is because nonidealities in later stages, when input-referred, are divided by the total gain preceding them. By design, each stage in a $\Delta\Sigma$M has a high gain in the signal band, so stages two and beyond have a reduced influence at the input, but the first stage does not. Thus, in a DT $\Delta\Sigma$M, considerable effort must be spent designing the first op amp [Yuk87] while requirements on later op amps may be relaxed.

Likewise, the first transconductor in a CT $\Delta\Sigma$M is the most important for overall thermal noise and linearity [Jen95, Mor98]. A mismatch in the input differential pair transistors leads to an offset which results in a dc term in the output spectrum; special care must be taken in converters where dc is not removed by the decimator.

COMPONENT NONLINEARITY

Earlier we mentioned nonlinearity in the first stage op amp gain characteristic adds harmonic distortion. The same thing happens if components near the input are nonlinear. For example, [Hau86] shows how a voltage-sensitive first integrating capacitor degrades performance in a typical SC $\Delta\Sigma$M.

In the CT $\Delta\Sigma$M of [Jen95], the input transconductor uses a differential pair degenerated with an emitter resistor to set g_m. The linearity of this resistor is the key to the linearity of the whole circuit. If the first integrating capacitor is slightly nonlinear, harmonics of the input signal appear in the output spectrum—the resulting spectrum looks similar to the one in Figure 3.2b. Usually, component linearity requirements are more stringent than component tolerance requirements.

CLOCK JITTER

How important is timing jitter in the quantizer clock in a $\Delta\Sigma$M? Compared to a Nyquist-rate converter, Harris [Har90] found $\Delta\Sigma$Ms had a tolerance to white jitter improved by the oversampling ratio for the same jitter variance. Boser [Bos88] found the same thing, but he also noted that because jitter noise falls as $1/\text{OSR}$ while quantization noise falls as $1/\text{OSR}^{2m+1}$, modulators with high OSRs are more likely to be performance-limited by jitter. Van der Zwan [vdZ96] presents an argument that CT $\Delta\Sigma$Ms are more sensitive to jitter than DT $\Delta\Sigma$Ms; we will take this issue up in Chapter 5 when we consider in detail the problem of clock jitter in CT $\Delta\Sigma$Ms.

2. A SURVEY OF IMPORTANT CT $\Delta\Sigma$M PAPERS

We very much enjoy papers in the literature which survey a particular area and include many of the important references in that area. We have performed just such a survey in the area of CT $\Delta\Sigma$M, and because we presume the reader feels the same way we do about lists of important references, we give a chronological listing of these papers with a brief description of each.

[Ino62], [Ino63] The first of these papers is where $\Delta\Sigma$M was first published, though there was a patent granted to Cutler a couple of years earlier [Cut60]. The second paper contains some analysis of a CT $\Delta\Sigma$M with both a single and double integrator. In particular, the 9dB and 15dB of SNR improvement per octave of oversampling for the first- and second-order modulators are derived. They build circuits for both and verify the predicted performance, and apply the circuits to the encoding of video signals.

Following this paper, over the next twenty years there were not many papers on $\Delta\Sigma$M because integrated MOS processes were still expensive. As they became cheaper, the DSP required in the decimator became cheaper, and hence $\Delta\Sigma$Ms began to receive more attention.

[Can85] It is this paper which really sparked interest in $\Delta\Sigma$M as a method for ADC. It is widely cited as the source for the double integration $\Delta\Sigma$M. As we have seen, such a $\Delta\Sigma$M contains two cascaded integrators and implements $\text{NTF}(z) = (1 - z^{-1})^2$, i.e., double differentiation of the quantization noise. We sometimes denote this modulator the *standard [low pass] second-order* $\Delta\Sigma$M since, as we shall see, there have been many subsequent implementations of it. This paper is the first to derive the DT/CT loop filter equivalence

$$H(z) = \frac{2z - 1}{(z - 1)^2} \quad \leftrightarrow \quad \hat{H}(s) = \frac{1 + 1.5sT_s}{s^2 T_s^2} \tag{3.2}$$

from (2.10) for a feedback DAC that emits NRZ pulses, and a CT $\Delta\Sigma$M circuit was built based on this equivalence.

[Koc86] This is also a double integration CT $\Delta\Sigma$M, following Candy's lead. With a 15MHz clock, the achieved performance was 77dB over a 120kHz bandwidth and a power consumption of 20mW—certainly not poor even by today's standards.

[Ada86] Miles ahead of its time, this paper describes a fourth-order four-bit (i.e., with a four-bit quantizer) CT $\Delta\Sigma$M that achieves 18 bits of resolution at 24kHz. The crucial issue of DAC waveform asymmetry (i.e., differing rise and fall times at the DAC output) is first discussed here.

[Pea87], [Sch89], [Gai89] The idea of band pass ΔΣM was, to our knowledge, first suggested in the earliest of these papers, though it can hardly be said that those proceedings are widely available. In *Electronics Letters*, Schreier and Snelgrove introduced the idea to a wider audience in the second paper listed; unbeknownst to them at the time, a U.S. patent had been granted to Gailus et al. a few months earlier, as the third citation shows.

[Gos88], [Gos90] These papers were the first to point out that a delay between the sampling clock edge and DAC pulse edge affects the performance of a CT ΔΣM. We have denoted this delay *excess loop delay* in Chapter 2 §4.2, and as we shall see in Chapter 4, it turns out to have a major impact on the design of CT ΔΣMs.

[Bro90] This was an early paper on how to simulate CT systems in DT. It used a third-order CT ΔΣM as an example and showed how to simulate its behavior both in C and SABER. The bilinear transform was used to map between z and s domains, though we prefer the impulse-invariant transformation.

[Hor90] Another paper very advanced for its time and often overlooked, it also discusses excess loop delay in CT ΔΣM and is the first to suggest the use of the modified $\mathcal{Z}$-transform to account for loop delay in the design of high-order CT ΔΣMs. While [Gos88] showed that a certain amount of loop delay is beneficial to a first-order CT ΔΣM, this paper extended the results to higher-order CT ΔΣMs.

[Asi91] Although no circuit was actually designed here, this is an early paper that considers some issues in the design of very fast CT ΔΣMs. For a standard second-order CT ΔΣM clocking at 500MHz in GaAs, the authors examine the effect on SNR of finite op amp dc gain, gain-bandwidth product, and signal swing, and small nonlinearities in the op amp, and conclude that a 10-bit converter could be built to work at this speed.

[Thu91] At the time this paper was published, the idea of band pass ΔΣM was relatively new. This paper is the first to use the impulse-invariant transformation to design a continuous-time BP ΔΣM. They designed a loop transfer function with nonoptimal noise shaping; it took [Sho94] to explain how to overcome this with the multi-feedback design from Chapter 2 §1.3.

[Cha92] The authors talk about design issues of a standard second-order CT ΔΣM in GaAs for 500MHz clocking, including finite op amp dc gain and gain-bandwidth product, and quantizer hysteresis and delay. They then fabricate and test a prototype whose poor performance is attributed to poor comparator sensitivity. However, the circuit was one of the first to demonstrate the feasibility of integrating high-speed CT $\Delta\Sigma$Ms.

[Can92b] This is the first of two IEEE Press books published about $\Delta\Sigma$M. It is a compendium of Candy and Temes' opinion of the important papers in $\Delta\Sigma$M up until early 1990. We recommend owning a copy to anyone working in $\Delta\Sigma$M design since most or all important early papers may be found in this convenient reference.

[Hal92] This is another early high-speed modulator, this time in 2μm CMOS clocking at 150MHz. It describes a standard second-order CT $\Delta\Sigma$M and achieves 10-bit resolution at an OSR of 128.

[Wan92] It had long been suspected that the standard double integration $\Delta\Sigma$M is stable for dc inputs up to the rails. This paper proves it using a geometrical argument about the bounds of internal states.

[Tro93] These authors designed a BP CT $\Delta\Sigma$M on an analog-digital FPGA. As in [Thu91], they didn't implement the correct $H(z)$.

[Sho94] It was in *this* paper that a correct method for designing band pass CT $\Delta\Sigma$Ms based on the impulse-invariant transform was explained.

[Ris94] Lars Risbo's doctoral thesis is unique. Motivated by his desire to build the best-sounding CD player possible, he examines stability and design methods for high-order single-bit $\Delta\Sigma$Ms in ways that are highly innovative and original. Sadly, his ideas will likely be appreciated by few because they are almost too clever: it took us three separate attempts over one year to grasp much of what Risbo says. Nonetheless, this is a reference work to be taken seriously for anyone wishing a deep understanding of $\Delta\Sigma$Ms. His Appendix C contains some discussion of clock jitter in CT $\Delta\Sigma$Ms, a topic we cover at length in Chapter 5.

[Fen94], [Nar94], [Jen94] Three high-speed CT $\Delta\Sigma$Ms appeared at the GaAs Integrated Circuits Symposium in 1994. The first listed clocked at 500MHz, the second at 2GHz, and the third at 4GHz; the first two are standard second-order low pass GaAs designs while the third is a first-order low pass InP design. All three designs suffered from moderate amounts of harmonic distortion in the baseband.

[Jen95] We spent considerable time studying this paper. It describes the building of a standard second-order CT $\Delta\Sigma$M using InP double heterostructure HEMTs clocking at 3.2GHz for converting a 50MHz baseband signal, and

71dB SFDR is achieved. A circuit diagram of their modulator was shown in Figure 1.15 on page 24.

[Mit95] The authors of this paper simulate, but don't build, a first-order CT ΔΣM in CMOS that dissipates only 3mW at a clock speed of 128MHz.

[Sho96], [Sho97] Omid Shoaei's excellent Ph.D. thesis is one of the definitive works on high-speed CT ΔΣM. Shoaei attempted to build a 250MHz fourth-order band pass CT ΔΣM for conversion of narrowband signals at 62.5MHz, but the final performance was thwarted by unexpectedly high fabrication tolerances and a lack of common-mode feedback circuitry in his transconductors. The first citation is the thesis itself, and the second is a journal paper which summarizes the thesis.

[Erb96] This brief paper describes a silicon bipolar standard second-order CT ΔΣM clocking at 1.28GHz. The performance is at best 8 bits, though the paper's length permits very little detail to be given. It is implied that the authors use a circuit architecture similar to that in Figure 1.15.

[Sch96a] This paper explains how to design a CT ΔΣM by transforming it to a DT ΔΣM design problem using the impulse-invariant transform. While [Sho96] deals with the problem in pole-zero form, [Sch96a] represents the modulator in state-space.

[Opa96], [Don97], [Don98b], [Don98a] Opal's 1996 work focuses on the rapid simulation of clocked CT systems in the DT domain. Since then, a student of his, Yikui Dong, has written several papers about rapidly simulating nonideal effects in CT ΔΣMs, most particularly thermal noise. Such a tool is extremely useful, and the simulation results look plausible, but unfortunately no experimental validation of their results has been provided to date.

[vdZ96], [vdZ97] These nice papers present fourth-order CT ΔΣMs with very low power. The first is a 0.2mW voice-band coder, the second a 2.3mW audio coder. Both achieve about 15 bits of performance. [vdZ96] also contains some discussion about tradeoffs between DT and CT designs.

[Ben97], [Gao97a] As noted, excess loop delay in a CT ΔΣM worsens performance. Both of these papers talk about how to compensate for loop delay with appropriate feedback and tuning. We will be examining this further in Chapter 4.

[Rag97] Written by the authors of [Jen95], this paper describes Figure 1.15 with an additional transconductor element to turn the low pass modulator into a mildly band pass one with a noise notch tunable from 0 to 70MHz and a 4GHz sampling clock. An SNR of 92dB was measured for a very narrow bandwidth corresponding to an OSR of around 5000.

[Che97] It is believed that this is the first mention of the performance effects of quantizer metastability in CT $\Delta\Sigma$Ms. It was written by us, and a large amount of material along these lines is given in Chapter 5 of this monograph.

[Nor97] This is the second IEEE Press book about $\Delta\Sigma$M. Instead of a compilation of papers, the editors commissioned various authors to write chapters in their areas of expertise on many aspects of $\Delta\Sigma$M theory, design, and simulation. It, too, is highly recommended as a reference work for anyone in the area—the book has acquired the moniker "the orange Bible".

[Miy97], [Olm98] In these papers, a 5GHz HEMT modulator was designed for a 50MHz bandwidth, and 7-bit performance was achieved. It must be noted, however, that this performance was achieved with a signal band that did not extend below 6MHz; this was apparently to avoid further SNR degradation by the $1/f$ noise of the devices.

[Jay97] Following [Jen95], these authors make a fourth-order BP CT $\Delta\Sigma$M clocking at 3.2GHz with an 800MHz center frequency. They achieve 7-bit performance in a 25MHz band, though they estimate with proper design this could be raised to 10 bits. We arrive at a similar conclusion for our 4GHz modulator in Chapter 6.

[Gao97b], [Gao98b] These papers present an aggressive 4GHz CT second-order band pass design in 25GHz Si bipolar technology for direct conversion of 950MHz analog signals. In a very narrow bandwidth, 10-bit performance was achieved.

[Che98a], [Che98b] We presented these papers at ISCAS 1998. The first deals with simulating CT $\Delta\Sigma$Ms, something we covered briefly in Chapter 1, and the second deals with the nonideal effects of excess loop delay, clock jitter, and quantizer metastability on the performance on CT $\Delta\Sigma$Ms. This monograph is in no small part an expansion of the second paper.

[Gao98a] Preliminary test results on a fourth-order 4GHz CT band pass modulator in a 40GHz SiGe HBT technology are presented in this paper. We greatly extend these results in Chapter 7.

[Oli98] In this paper the effect of jitter in the DAC pulse width of return-to-zero-style first- and second-order modulators is studied. The authors conclude, as [Che97] did earlier, that jitter in the width of the DAC pulse is not noise-shaped and hence degrades performance. However, their new result is that a second-order modulator provides first-order shaping of pulse *starting time* jitter; hence, they propose using a monostable multivibrator as a quantizer, which produces fixed pulse widths even in the presence of variable pulse start time.

[Mor98] This is the first paper to our knowledge which contains a high-speed design of order three. This circuit contains two separate modulators for I and Q channels with integrated mixers, similar in architecture to Figure 1.7 only where the mixing is done as part of the first stage of the modulator so that the modulators themselves are low pass. In a 50MHz bandwidth, 35dB SNR was achieved.

3. HIGH-SPEED CT ΔΣM PERFORMANCE

It is interesting to see how the published high-speed designs fare in terms of performance. Table 3.1 summarizes the order and type of the high-speed CT ΔΣM designs surveyed, where "1LP" means first-order low pass, "2BP" means second-order band pass, etc. The majority of the designs are implementations of the standard double integration modulator. The two second-order BP modulators are $f_s/4$ modulators which ideally have the same performance and stability as a first-order LP design. The two fourth-order BP designs are also $f_s/4$ designs, and they have the same performance and stability as a double integration ΔΣM. Thus, all the high-speed designs listed except the last one are first- or second-order. For each clock rate f_s and OSR, DR and SNR_{max} are also listed.

The performance of an ideal first- or second-order modulator can be found from DT simulation as we did in Example 1.5 or from [Sch93, Fig. 7]. Based on these, an approximate formula for the achievable performance in a double integration modulator is

$$\text{DR} \approx 15\log_2(\text{OSR}) - 13\text{ dB}, \ \text{SNR}_{\text{max}} \approx 15\log_2(\text{OSR}) - 20\text{ dB}. \qquad (3.3)$$

Table 3.2 shows how each of the published 2LP/4BP modulators compares to (3.3). Generally, we see performance falling far short of ideal, particularly for OSRs of 64 or more. Frequently in papers that publish output spectra it is clear that the signal band is filled with white, rather than shaped, noise. Thus, doubling the OSR results only in a 3dB DR improvement instead of 15dB for an ideal second order ΔΣM as depicted in Figure 3.3. We could extrapolate backwards to the approximate point where noise shaping ends and white noise begins as

$$\text{Useful OSR} \approx \text{OSR} \div 2^{\text{DR loss}/12}. \qquad (3.4)$$

This is termed "useful OSR" because it is the OSR for which noise shaping ceases, and it is listed in the table for each modulator.

Table 3.1. High-speed CT ΔΣM published performance.

Paper	Technology	Type	f_s	OSR	DR (dB)	SNR (dB)
[Cha92]	1μm GaAs D MESFET	2LP	200MHz	100	58	50
[Hal92]	2μm CMOS	2LP	150MHz	128	63	57
[Fen94]	0.5μm GaAs HEMT	2LP	500MHz	100	60	55
[Nar94]	1.4μm GaAs HBT	2LP	2GHz	20	43	37
[Jen95]	2.0μm InP DHBT	2LP	3.2GHz	32	49	50
[Mit95]	1.2μm CMOS	1LP	128MHz	128	60	57
[Sho96]	0.8μm Si BiCMOS	2BP	200MHz	500	50	46
[Erb96]	Si BJT	2LP	1.28GHz	64	–	45
[Rag97]	2.0μm InP DHBT	2LP/BP	4GHz	64	44	41
[Gao97b]	0.5μm Si BJT	2BP	3.8GHz	10000	60?	49
[Jay97]	AlGaAs/GaAs HBT	4BP	3.2GHz	64	42	41
[Olm98]	0.4μm InGaP/InGaAs HEMT	2LP	5GHz	50	51	39
[Gao98a]	0.5μm SiGe HBT	4BP	4GHz	500	62	53
[Mor98]	0.5μm SiGe HBT	3LP	1.6GHz	16	–	35

Table 3.2. Performance in published double integration CTΔΣMs relative to ideal simulation.

Modulator	[Cha92]	[Hal92]	[Fen94]	[Nar94]	[Jen95]
f_s (GHz)	0.2	0.15	0.5	2.0	3.2
OSR	100	128	100	20	32
DR loss (dB)	29	29	27	9	13
Useful OSR	18.7	24.0	21.0	11.9	15.1
Modulator	[Erb96]	[Rag97]	[Jay97]	[Olm98]	[Gao98a]
f_s (GHz)	1.28	4.0	3.2	5.0	4.0
OSR	64	64	64	50	500
DR loss (dB)	25?	33	35	21	59
Useful OSR	15.1	9.5	8.5	14.9	16.6

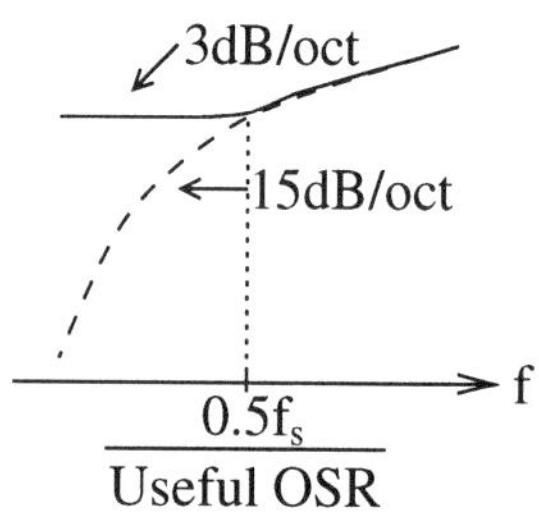

Figure 3.3. Ideal vs. real spectra in double integration modulators.

Clearly, there is little benefit in using OSR $>$ 15 for GHz-speed modulators. It is surprising how consistent this number is, even with quite different clock speeds and semiconductor processes. The problem is not with CT ΔΣMs in general—[Ada86], for example, achieved DR $=$ 105dB in a 20kHz band—it is with *high speed* CT ΔΣMs. It might be that all these modulators are limited simply by thermal noise; the same thermal noise spec would cause 30dB more noise in a 20MHz band than in a 20kHz band, so a 4GHz modulator would be more likely to be thermal-noise limited than a 4MHz modulator with the same OSR. Still, there are a number of other possibilities which we explore in the coming chapters.

4. SUMMARY

Unsurprisingly, a number of the nonidealities in DT modulators affect CT modulators too. As far as published high-speed CT ΔΣMs go, they achieve poor performance compared to an ideal modulator. We spend the the next two chapters investigating the major performance-limiting nonidealities in high-speed CT ΔΣMs.

Chapter 4

EXCESS LOOP DELAY

Consider once again the high-speed double integration modulator from Figure 1.15 [Jen95] reproduced in Figure 4.1. The quantizer is a latched comparator

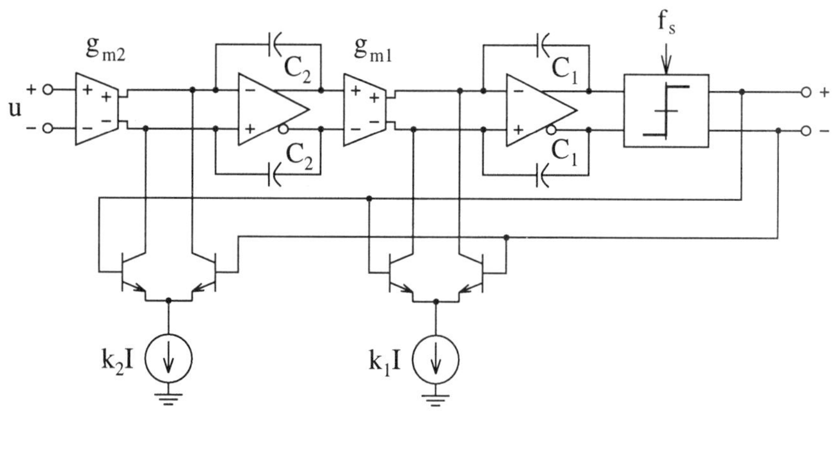

Figure 4.1. Example of high-speed double integration CT ΔΣM.

whose output drives differential pair DACs; their output currents sum with the transconductor outputs. Ideally, the DAC currents respond immediately to the quantizer clock edge, but in practice, the transistors in the latch and the DAC cannot switch instantaneously. Thus, there exists a delay between the quantizer clock and DAC current pulse. It is this delay that we called *excess loop delay* in Chapter 2 §4.2; we also call it simply *excess delay* or *loop delay*. There is really no analogous problem in DT modulators; perhaps the closest thing is incomplete settling, which we discussed in Chapter 3 §1.1.

Excess delay has been studied in the literature before; a brief summary of past work is appropriate. Gosslau and Gottwald [Gos88, Gos90] found that excess delay of 25% actually improves the DR of a 1LP CT $\Delta\Sigma$M. Horbach [Hor90] confirmed this and extended the results to higher-order LP modulators, showing that excess delay is detrimental to their performance. Chan [Cha92] found a full sample of feedback delay in his 2LP modulator caused 10dB of SNR loss. Shoaei [Sho96] found excess delay problematic in 2BP and 4BP modulators. Gao et al. [Gao97a] propose feedback coefficient tuning and demonstrate that it alleviates delay problems in a 4BP modulator, while Benabes et al. [Ben97] add an extra feedback loop to a 2LP modulator for the same purpose.

In this chapter we unify and summarize the past work in the area, as well as contribute new material. First, most authors use the modified $\mathcal{Z}$-transform for studying excess delay, but we explain here why this is inappropriate and demonstrate a preferred method. Second, we consider higher-order LP and BP modulators in much more detail than has previously appeared, and treat modulators with a multibit quantizers for the first time. Our own published work on this subject has appeared in [Che98b] and [Che99b].

1. EFFECT OF EXCESS LOOP DELAY

As noted in the introduction, excess loop delay arises because of nonzero transistor switching time, which makes the edge of the DAC pulse begin *after* the sampling clock edge. For an NRZ DAC pulse, this is depicted in Figure 4.2. The sampling instant is $t = 0$.

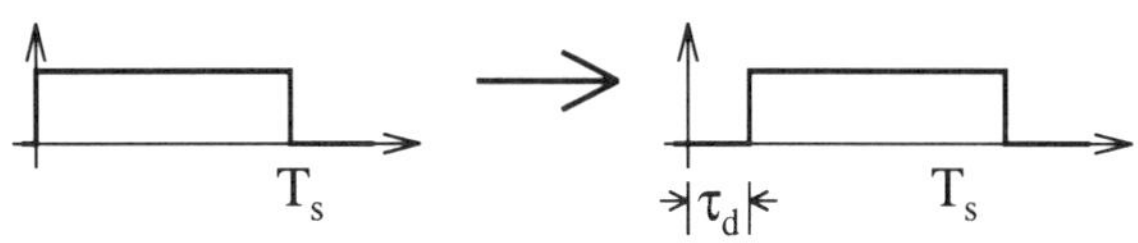

Figure 4.2. Illustration of excess loop delay on NRZ DAC pulse.

We assume that excess loop delay can be expressed as a fraction of the sampling period

$$\tau_d = \rho_d T_s. \tag{4.1}$$

In an actual circuit, the value of τ_d depends on the switching speed of the transistors f_T, the quantizer clock frequency f_s, and the number of transistors in the feedback path n_t (as well as other things like the loading on each transistor). As a crude approximation, we could assume transistors switch fully at the maximum speed, i.e., after time $1/f_T$, in which case we could write

$$\rho_d \approx \frac{n_t f_s}{f_T}. \tag{4.2}$$

τ_d could end up being a significant fraction of T_s depending on the parameters in (4.2). This is particularly likely in GHz-speed modulators built in a process with an f_T of a few tens of GHz.

Example 4.1: In the design in Figure 4.1, suppose we desire 12-bit DR in a 50MHz bandwidth. This will require an OSR of about 50 [Sch93], which means we must clock at $f_s = 50(2 \cdot 50) = 5\text{GHz}$. If the quantizer is an ECL-style latched comparator, its output differential pair must switch; the DAC must also switch, and thus $n_t = 2$. In a $f_T = 30\text{GHz}$ process, therefore,

$$\rho_d \approx \frac{2 \cdot 5}{30} = 33\% \tag{4.3}$$

is the amount of excess delay predicted by (4.2). □

The reason why excess loop delay is problematic can be understood by recalling the CT/DT equivalence from Chapter 2 §1.1. Recall that we were treating DAC pulses as rectangular with the form

$$\hat{r}_{(\alpha,\beta)}(t) = \begin{cases} 1, & \alpha \le t < \beta,\ 0 \le \alpha < \beta \le 1 \\ 0, & \text{otherwise.} \end{cases} \tag{4.4}$$

in (2.4). Suppose we have assumed that we have an NRZ DAC with $(\alpha, \beta) = (0, 1)$, and we have found the equivalent $\hat{H}(s)$ for a desired $H(z)$ using Table 2.1 or Matlab. If we then go and use this filter in a system with delayed pulses as in Figure 4.2, then the system no longer has the same α and β. This means the equivalence between $\hat{H}(s)$ and $H(z)$ is affected.

We can calculate the effect of loop delay mathematically by using Table 4.1, which lists the z-domain equivalents for s-domain $\hat{H}(s)$ poles of orders one through three. As with Table 2.1, these were calculated with the help of Maple and (2.2). An s-domain pole of multiplicity l at s_k maps to one at z_k with the same multiplicity, with

$$z_k = \exp s_k. \tag{4.5}$$

Poles at $s_k = 0$ give numerators of $0^l/0^l$, as before, and the rightmost column gives the formulae that result when l'Hôpital's rule is applied l times. Of course, we could also use Matlab to do the transformations.

Example 4.2: Suppose we are designing a CT double integration modulator. If we had NRZ DAC pulses, then we would have found $\hat{H}(s)$ to be

$$\hat{H}(s) = -\frac{1 + 1.5s}{s^2} \tag{4.6}$$

from (2.10). Suppose further that we have excess loop delay τ_d, so that in actuality we have NRZ DAC pulses delayed by τ_d as in Figure 4.2. In that

Table 4.1. z-domain equivalences for s-domain loop filter poles.

s-domain pole	z-domain equivalent	Limit for $s_k = 0$
$\frac{r_0}{s-s_k}$	$\frac{y_0}{z-z_k} \times \frac{r_0}{s_k}$ $y_0 = z_k^{1-\alpha} - z_k^{1-\beta}$	$\frac{y_0}{z-z_k}$ $y_0 = r_0(\beta - \alpha)$
$\frac{r_0}{(s-s_k)^2}$	$\frac{y_1 z + y_0}{(z-z_k)^2} \times \frac{r_0}{s_k^2}$ $\begin{aligned} y_1 &= z_k^{1-\beta}[1 - s_k(1-\beta)] \\ &- z_k^{1-\alpha}[1 - s_k(1-\alpha)] \\ y_0 &= z_k^{2-\alpha}(1 + s_k\alpha) \\ &- z_k^{2-\beta}(1 + s_k\beta) \end{aligned}$	$\frac{y_1 z + y_0}{(z-z_k)^2}$ $\begin{aligned} y_1 &= \tfrac{r_0}{2}[\beta(2-\beta) - \alpha(2-\alpha)] \\ y_0 &= \tfrac{r_0}{2}(\beta^2 - \alpha^2) \end{aligned}$
$\frac{r_0}{(s-s_k)^3}$	$\frac{y_2 z^2 + y_1 z + y_0}{(z-z_k)^3} \times \frac{r_0}{s_k^3}$ $\begin{aligned} y_2 &= z_k^{1-\beta}[-1 + s_k(1-\beta) + \tfrac{s_k^2}{2}(1-\beta)^2] \\ &- z_k^{1-\alpha}[-1 + s_k(1-\alpha) + \tfrac{s_k^2}{2}(1-\alpha)^2] \\ y_1 &= z_k^{2-\beta}[2 - s_k(1-2\beta) \\ &+ \tfrac{s_k^2}{2}(-1 - 2\beta + 2\beta^2)] \\ &+ z_k^{2-\alpha}[2 - s_k(1-2\alpha) \\ &+ \tfrac{s_k^2}{2}(-1 - 2\alpha + 2\alpha^2)] \\ y_0 &= z_k^{3-\alpha}(1 + s_k\alpha + \tfrac{s_k^2}{2}\alpha^2) \\ &- z_k^{3-\beta}(1 + s_k\beta + \tfrac{s_k^2}{2}\beta^2) \end{aligned}$	$\frac{y_2 z^2 + y_1 z + y_0}{(z-z_k)^3}$ $\begin{aligned} y_2 &= r_0[\tfrac{1}{6}(\beta^3 - \alpha^3) \\ &- \tfrac{1}{2}(\beta^2 - \alpha^2) + \tfrac{1}{2}(\beta - \alpha)] \\ y_1 &= r_0[\tfrac{1}{3}(\beta^3 - \alpha^3) \\ &- \tfrac{1}{2}(\beta^2 - \alpha^2) - \tfrac{1}{2}(\beta - \alpha)] \\ y_0 &= \tfrac{r_0}{6}(\beta^3 - \alpha^3) \end{aligned}$

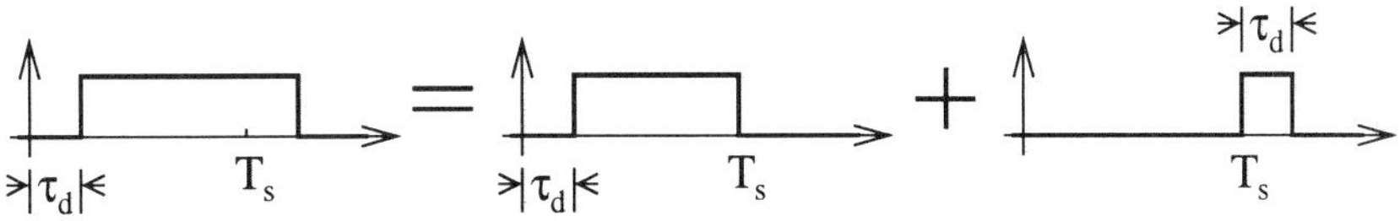

Figure 4.3. Delayed NRZ pulse as a linear combination.

case, we have $(\alpha, \beta) = (\tau_d, 1 + \tau_d)$. What is the equivalent $H(z)$ for such an $\hat{H}(s)$ and DAC pulse?

The formulae in Table 4.1 only apply for a pulse with $\beta \le 1$, but once again, superposition comes to our rescue: it is possible to write a τ_d-delayed NRZ pulse as

$$\hat{r}_{(\tau_d, 1+\tau_d)}(t) = \hat{r}_{(\tau_d, 1)}(t) + \hat{r}_{(0,\tau_d)}(t-1), \tag{4.7}$$

that is, as a linear combination of a DAC pulse from τ_d to 1 *and* a one-sample-delayed DAC pulse from 0 to τ_d as shown in Figure 4.3. Writing (4.6) in partial fractions gives

$$\hat{H}(s) = \frac{-1.5}{s} + \frac{-1}{s^2}. \tag{4.8}$$

Applying Table 4.1 to each term of (4.8), for each of the two DAC pulses in (4.7), yields

$$\frac{-1.5}{s} \rightarrow \frac{-1.5(1-\tau_d)}{z-1} + z^{-1}\frac{-1.5\tau_d}{z-1} \tag{4.9}$$

$$\begin{aligned} \frac{-1}{s^2} &\rightarrow \frac{(-0.5+\tau_d-0.5\tau_d^2)z + 0.5(-1+\tau_d^2)}{(z-1)^2} \\ &+ \quad z^{-1}\frac{\tau_d(-1+0.5\tau_d)z - 0.5\tau_d^2}{(z-1)^2}. \end{aligned} \tag{4.10}$$

Adding (4.9) and (4.10) gives

$$H(z,\tau_d) = \frac{(-2+2.5\tau_d-0.5\tau_d^2)z^2 + (1-4\tau_d+\tau_d^2)z + (1.5\tau_d - 0.5\tau_d^2)}{z(z-1)^2}. \tag{4.11}$$

We can quickly verify that for $\tau_d = 0$, (4.11) turns into (2.7),

$$H(z) = \frac{-2z+1}{(z-1)^2}, \tag{4.12}$$

as it should. However, for $\tau_d \neq 0$, the equivalent $H(z)$ is no longer (4.12).

If instead of Table 4.1 we use the modified $\mathcal{Z}$-transform on (4.12), the result is [Gao97a]

$$\begin{aligned} H_{MZ}(z, \tau_d) &= \mathcal{Z}_m \left[\mathcal{Z}^{-1} \left[\frac{-2z+1}{(z-1)^2} \right] \right] \\ &= \frac{(-2+2\tau_d)z^2 + (1-3\tau_d)z + \tau_d}{z(z-1)^2}, \end{aligned} \tag{4.13}$$

which is similar to (4.11) but not identical. The modified $\mathcal{Z}$-transform assumes the delay happens at the output of $\hat{H}(s)$ (at the quantizer *input*), but in our method we assume the delay happens prior to the DAC pulse (at the quantizer *output*). Papers in the literature generally do not distinguish between these two cases, but they are clearly different.

We believe our assumption is superior because it represents reality more closely—the delay is after the quantizer in an actual circuit—so we prefer to use partial fractions and Table 4.1 rather than the modified $\mathcal{Z}$-transform to compute the effect of excess delay. □

As was the case in Chapter 2 §1.2, we treat pulses as rectangular because it allows exact closed-form solutions in the CT/DT equivalence calculations. Other authors [Ben97] treat pulses as trapezoidal or as having exponential rising behavior, which is more realistic but which does not lead to exact solutions as easily. As we have said, a real circuit will likely exhibit DAC pulse shapes more complicated still. The key point is, excess delay always alters the numerator coefficients of the equivalent $H(z)$, and it turns out that using rectangular pulses yields simulation results that are similar to those found using more realistic pulse shapes.

To study the effects of excess loop delay, Matlab code was written to perform the transformations in Tables 2.1 and 4.1 numerically. The output bit stream from a modulator was determined by solving the difference equation $X(z) = G(z)H(z)U(z) + H(z)Y(z)$ in the time domain with a C program for given $G(z)$ and $H(z)$. The virtue of using the transformations is that it allows us to simulate in the DT domain, a process usually significantly more rapid than simulating using $\hat{H}(s)$ in the CT domain. Though we provide no experimental verification of the results throughout this chapter, we have found simulation of Figure 4.1 in Simulink using ideal circuit components and a variable delay in the feedback path gives results that are consistent with those presented here. Simulations take much longer with Simulink, however.

2. DOUBLE INTEGRATION MODULATOR

Since first-order modulators with excess delay have been studied already [Gos88] and are of limited practical use due to an excessive presence of harmonics in the output spectrum, we confine ourselves to modulators of orders two and above. In this section, we commence with the double integration $\Delta\Sigma$M in (4.11): how is its DR affected by excess delay? We said in Chapter 1 §3.3 that DR is defined as the difference between the smallest and largest input levels (in dB) which give SNR ≥ 0. At low input levels, SNR is limited by in-band quantization noise (IBN), while a large-enough input level eventually compromises the stability of the modulator. There exists a maximum stable input amplitude (MSA); DR may be found from IBN and MSA, as we explain below.

2.1 ROOT LOCUS

The easiest way to grasp the effect of excess delay is to linearize the quantizer as was done in Chapter 1 §1.1 and look at the stability of the noise transfer function. There is, however, a subtlety we ignored in Figure 1.1: the gain of a one-bit quantizer isn't well-defined. That is, we could insert a positive gain κ immediately in front of the quantizer and not affect the performance of the circuit—quantizer inputs would be scaled, but their signs remain unchanged, hence the sequence of ± 1 would be identical. Making κ explicit is usually done (Figure 4.4) in the linear model, which results in $\text{NTF}(z, \tau_d) = (1 +$

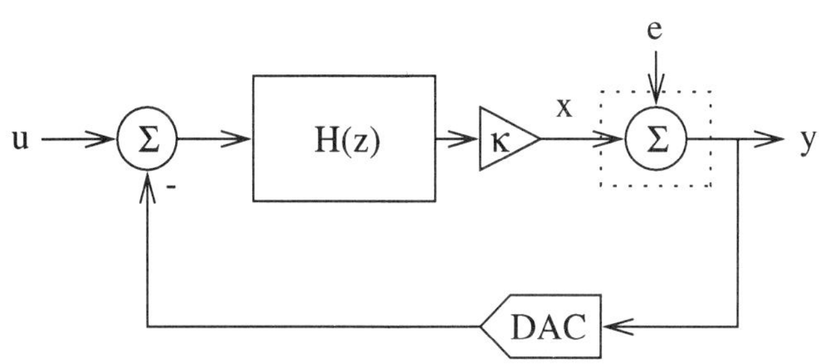

Figure 4.4. Linearized $\Delta\Sigma$M with one-bit quantizer arbitrary gain κ.

$\kappa H(z, \tau_d))^{-1}$. Figure 4.5 shows that for $\kappa = 1$ and increasing ρ_d, the poles of $\text{NTF}(z, \tau_d)$ move towards the unit circle, eventually moving outside at $\rho_d \approx 0.31$. Any choice of $\kappa > 0$ shows a similar movement of poles from their initial positions towards the unit circle; this implies modulator stability worsens as delay increases.

Time-domain simulation shows that it takes $\rho_d \approx 0.65$ to make the modulator unstable. The root locus incorrectly predicted $\rho_d = 0.31$ for instability because $\kappa = 1$ was not satisfied in the simulation. How to measure or choose κ is a nontrivial matter [Ris94, Chap. 6] and we do not explore it in detail here; we attempted to use the linear model for characterizing delay with little success. Suffice it to say that it at least makes a *qualitative* prediction that stability worsens with increasing loop delay.

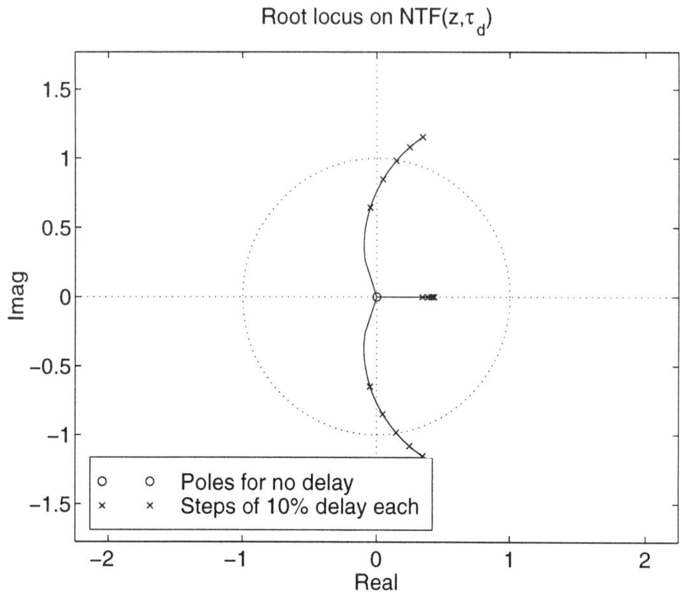

Figure 4.5. Effect of loop delay on root locus of NTF(z, τ_d).

2.2 IN-BAND NOISE

In practical terms, we care about how much performance is lost due to excess delay. Figure 4.6a shows an output spectrum near dc: 256 16384-point Hann-windowed periodograms with random initial conditions were averaged, and the input signal is a 0.1V sine wave. As the delay increases from 0% up to 60%, we see that the noise floor rises slowly. Integrating the IBN for zero input as a function of τ_d produces Figure 4.6b: for delays below about 20%, IBN stays roughly constant, but rises as delay increases[1]. If the excess delay exceeds about 65%, the modulator goes unstable. In this chapter, instability is defined to have occurred if the quantizer input magnitude exceeds 10 before the end of a simulation for 1000 successive simulations with random initial states. A similar definition was used in [Ris94].

The smallest input signal for which SNR $=$ 0dB is exactly the IBN, adjusted for the gain of the window (0.375 for Hann, or 4.26dB) and the fact that periodograms measure rms power (3.01dB). For example, the IBN for $\tau_d = 0$ and OSR $= 64$ is -85.06dB, and so we predict that an input magnitude of

[1]The non-monotonicity on the tails of the IBN graphs and certain later DR graphs is not a real effect: it is an artifact of doing simulations with zero input and no dither. Otherwise, the general trends indicated by the curves are accurate.

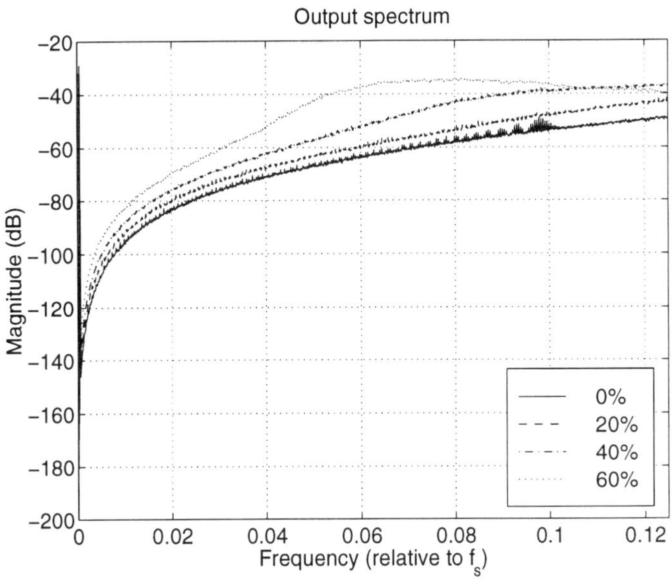

Figure 4.6a. Output spectrum from double integration CT ΔΣM as a function of excess delay.

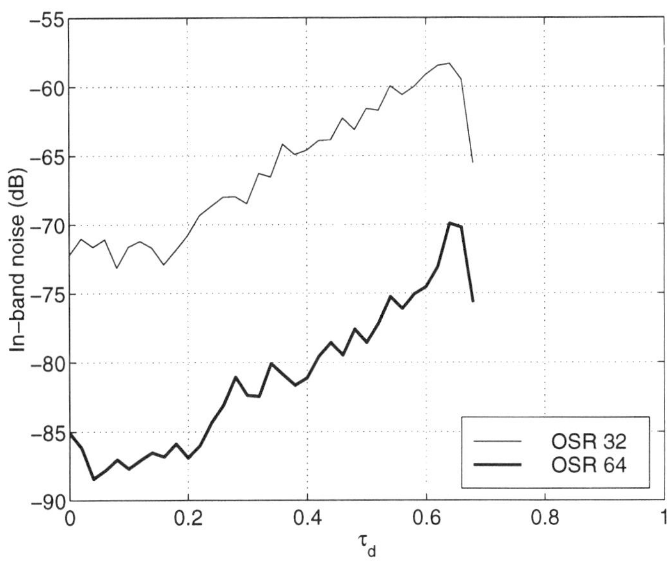

Figure 4.6b. In-band noise for zero input as a function of excess delay.

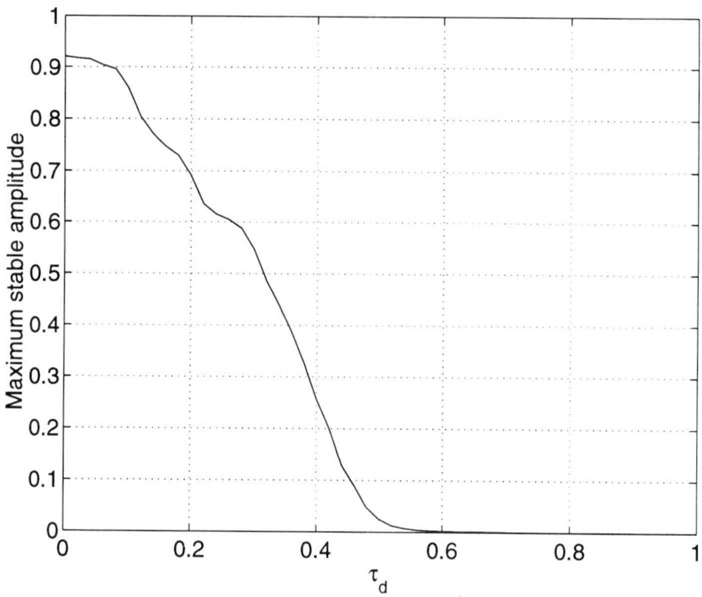

Figure 4.7. Maximum stable amplitude for double integration CT $\Delta\Sigma$M.

approximately

$$-85.06 + 4.26 + 3.01 = -77.79\text{dB} \tag{4.14}$$

is needed to get SNR $=$ 0dB. In simulation, we find the input magnitude that leads to 0dB SNR to be about -77dB.

2.3 MAXIMUM STABLE AMPLITUDE

The second question is, how does the MSA change with loop delay? To determine the MSA, we once again follow [Ris94]: we apply a ramp input whose amplitude increases slowly from 0 to 1 over 10^5 time steps. When the quantizer input magnitude exceeds 10, the input level at that instant is the MSA. A traditional method [Sch93] involves applying a low-frequency sine wave at the input and running for hundreds of thousands of cycles to check that the modulator remains stable, then increasing the amplitude and repeating the simulation until the maximum amplitude for which the modulator remains stable is found. We find Risbo's method gives approximately the same answer while requiring a small fraction of the number of simulation runs.

Performing this test for 200 runs with random initial conditions yields the graph in Figure 4.7. The modulator is stable for inputs of up to 0.92 for no excess delay, but this falls more or less linearly to near zero at about 50% delay.

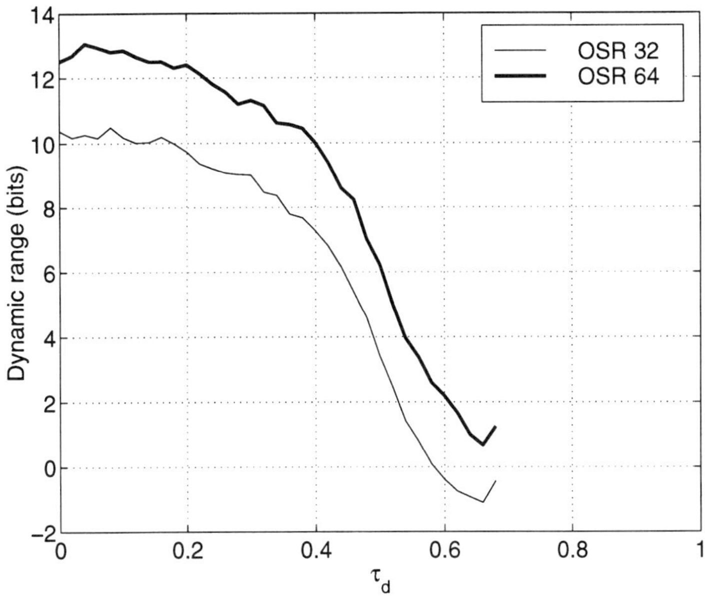

Figure 4.8. Dynamic range for double integration CT ΔΣM.

An unstable modulator has SNR $= -\infty$, so the MSA is precisely the largest input for which SNR > 0. For example, at $\tau_d = 0$, the MSA is

$$20 \log_{10} 0.92 = -0.72\text{dB}. \tag{4.15}$$

2.4 DYNAMIC RANGE

We can combine the previous two results to plot the modulator's DR against delay in Figure 4.8. DR is the difference between MSA and adjusted IBN; for example, at $\tau_d = 0$, equations (4.14) and (4.15) give

$$\text{DR} = -0.72 - (-77.79) = 77.07\text{dB}. \tag{4.16}$$

This is converted to bits using (1.10) and the result is plotted for $0 \leq \tau_d \leq 1$ in Figure 4.8.

Example 4.3: Example 4.1 estimated a loop delay of 33% in (4.3) for OSR $= 50$. We see from Figure 4.8 that even with OSR $= 64$ it would not be possible to achieve the desired resolution at 33% delay: we could only obtain DR $= 11$ bits. To achieve 12 bits at OSR $= 64$, we must have no more than about 20% excess loop delay. For a 50MHz bandwidth, OSR of 64 means clocking at 6.4GHz, and from (4.2), we see that the transistors must have $f_T > 32$GHz or so. □

3. $F_S/4$ FOURTH-ORDER BAND PASS MODULATOR

How does excess delay affect the design from (2.15) in Chapter 2 §1.3? Because we recommended using the multi-feedback architecture for BP modulators, we now have two DACs, both of whom have their leading pulse edge delayed by τ_d.

Exactly the same simulations were carried out for this BP modulator as were done in the previous section (IBN and MSA), only instead of using a dc input to find the MSA, a sine wave input at $f_s/4$ whose amplitude increases from 0 to 1 over 10^5 time steps is used. Again, this method is rapid, and we find it gives similar results to using a sine wave input with fixed amplitudes and frequencies near $f_s/4$, simulating for many cycles to see if the modulator remains stable, then increasing the amplitude and repeating the simulation.

The resulting DR as a function of τ_d is plotted in Figure 4.9a; for comparison, the results from Figure 4.8 for the double integration modulator are overlaid with dashed lines. Interestingly, the two designs have the same DR until about about 30% excess delay, at which point the BP design becomes more severely affected; the exact reason for this is unclear to the author. It goes unstable for about 50% excess delay. This particular example had NRZ and RZ DAC pulses, but the results do not change if a different pair of DAC pulses are selected.

For interest's sake, Figure 4.9b plots the same results only with OSR as the independent variable; the parameter on the curves is the product of τ_d and f_N, the Nyquist rate. Thus, for example, a modulator with a desired $f_N = 2$MHz and a fixed delay of $\tau_d = 1$ns has $\tau_d \times f_N = 10^{-3} = 0.1\%$, and the DR at a given clock speed $f_s \equiv \text{OSR} \cdot f_N$ may be found from the graph. Such a graph is useful when the technology and desired conversion bandwidth are known, and hence the excess delay calculated and the required clock speed for a given resolution found—we might wish to know how fast we must clock to achieve a given resolution specification.

Previous examinations of this modulator [Sho96, §3.1.4], [Gao97a] which found 25% delay required for instability made two errors. First, the modified $\mathcal{Z}$-transform was used which led to an incorrect $H_{BP}(z, \tau_d)$. Second, simulations were carried out with a large fixed-amplitude tone, which fails to take into account the changing modulator MSA with increasing delay.

4. HIGHER-ORDER MODULATORS

We now turn to the effects of excess loop delay for low pass CT ΔΣMs of order higher than two. This has been examined cursorily using the modified

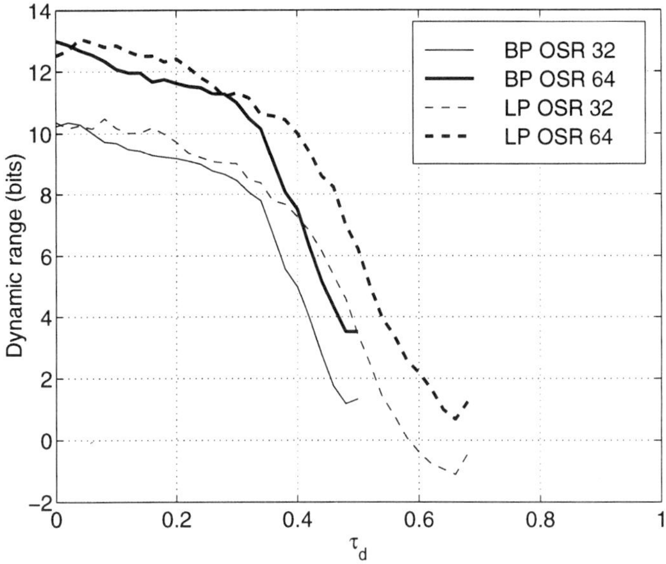

Figure 4.9a. Dynamic range for multi-feedback BP CT ΔΣM with comparison to double integration results.

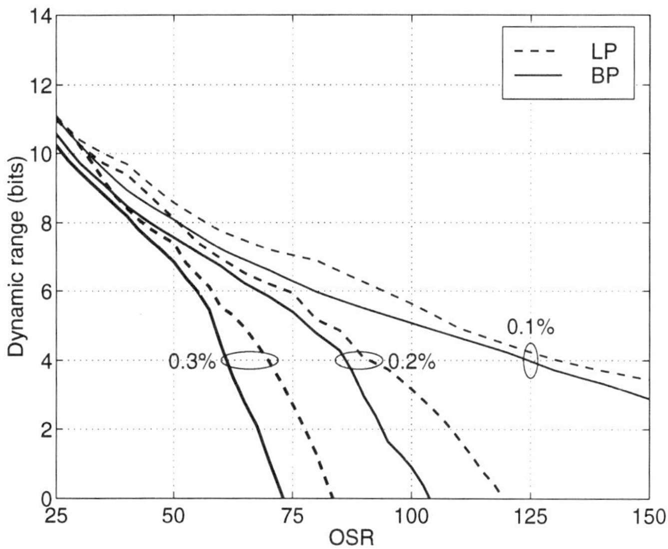

Figure 4.9b. Same graph with OSR as independent variable; numbers on curves are $\tau_d \times f_N$.

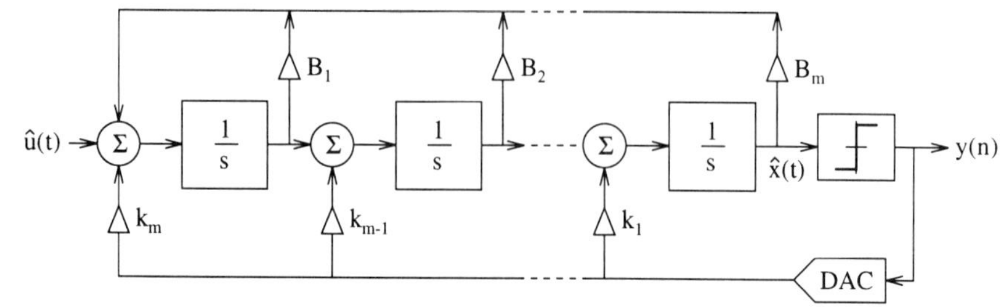

Figure 4.10. Block diagram for general high-order LP CT ΔΣM.

$\mathcal{Z}$-transform in [Hor90], but nowhere else to our knowledge. The architecture we will consider is a generalization of Figure 2.2 which was first shown in Figure 1.6, reproduced here with CT integrators in Figure 4.10. It is straightforwardly realizable in fully-integrated form with transconductors, integrators, and differential pair DACs as in Figure 4.1. The loop filter realized by this architecture for $m \geq 2$ is

$$\hat{H}(s) = \frac{\sum_{i=1}^{m} (\frac{1}{s})^i [k_i - \sum_{j=1}^{i-1} k_{i-j} B_j]}{1 - \sum_{i=1}^{m} B_i (\frac{1}{s})^i}. \tag{4.17}$$

(4.17) shows that the purpose of the B_is is to allow us to implement NTF(z) zeros at places other than dc (i.e., $z = 1$). Spreading the zeros out inside the band can result in significant SNR gains [Nor97, Chap. 4].

Four types of high-order modulators were designed using NTF prototyping. The NTFs used had

- Third-order Butterworth poles, all zeros at $z = 1$;
- Third-order Butterworth poles, optimally-spread zeros;
- Fourth-order Butterworth poles, optimally-spread zeros;
- Fifth-order Chebyshev poles, optimally-spread zeros.

The spread-zero modulators had zeros placed according to [Sch93] so that IBN would be minimized for a given OSR. Modulators with out-of-band gains (OOBGs) of 1.3, 1.4, 1.5, and 1.6 were all designed. For a few years now it has been recognized that choosing OOBG means a modulator with lower IBN, but at the price of lower MSA [Nor97, Chap. 4].

Example 4.4: We demonstrate this quickly for a fifth-order Chebyshev modulator with zeros spread assuming OSR $= 64$. Figure 4.11a shows that IBN falls from -116dB for OOBG $= 1.3$ down to -131dB for OOBG $=$ 1.6. At the same time, MSA in Figure 4.11b falls from about 0.79 to about

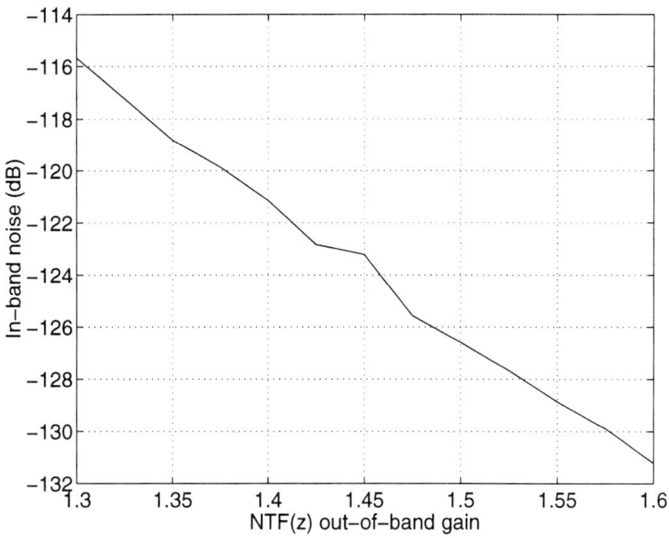

Figure 4.11a. Effect of out-of-band gain on fifth-order Chebyshev modulator in-band noise.

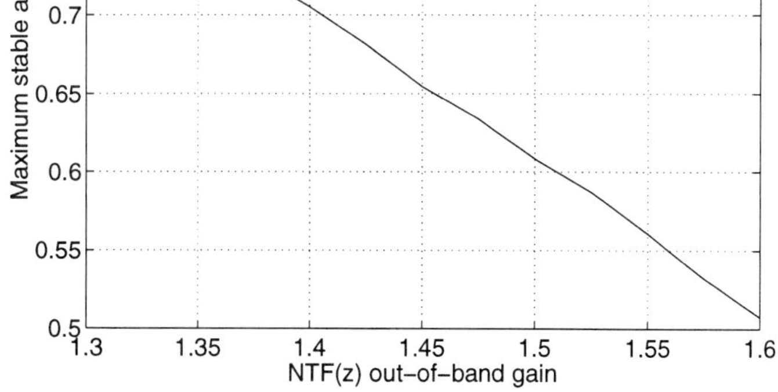

Figure 4.11b. Effect of out-of-band gain on fifth-order Chebyshev modulator maximum stable amplitude.

0.51. Note that IBN is plotted on a logarithmic scale but MSA is on a linear scale; DR increases from 17.4 to 19.3 bits over that range of OOBG. □

The DR as a function of excess loop delay for NRZ DAC pulses and OSRs of both 32 and 64 are summarized in the graphs in Figure 4.12. The results are most intriguing. The modulators with OOBG $= 1.3$ remain stable even for one full sample excess delay, and moreover they only suffer a dynamic range loss of between two and three bits. This contrasts starkly with the results for the second-order LP and fourth-order BP circuits. Increasing OOBG results in modulators which have generally better resolution at no delay, but which become unstable for less excess delay. This makes intuitive sense: higher OOBG means a generally less-stable $H(z)$, and in fact we see the τ_d needed for instability is roughly inversely proportional to OOBG. This suggests that higher-order modulators enjoy an advantage over the lower-order ones: the existence of a parameter, OOBG, which we may select according to our resolution *and* excess delay imperviousness requirements, though only extremely recently has anyone published a high-speed CT $\Delta\Sigma$M design with an order higher than two (recall [Mor98] in Table 3.1). To be fair, one *can* vary the OOBG in a second-order LP $\Delta\Sigma$M, but it is rarely done in practice.

For interest's sake, a sixth-order $f_s/4$ BP design was also tested by taking the low pass NTF with third-order Butterworth poles and three dc zeros and transforming it to a band pass design using $z^{-1} \to -z^{-2}$. This can be implemented using the multi-feedback architecture in Figure 2.4 with a third resonator and an additional feedback coefficient for each DAC. DR is plotted against τ_d in Figure 4.13. Comparing these curves to those of the equivalent third-order LP design in Figure 4.12 illustrates behavior like that in Figure 4.9a: the BP curves have the same shape as those of the LP curves for low excess delay, but they become unstable sooner as excess delay increases. Significantly, the LP modulator with OOBG $= 1.3$ was stable for a full sample of excess delay, while the same BP modulator is only stable up until $\tau_d = 0.65$.

In conclusion, LP modulators of order higher than two let us choose OOBG as an anti-delay measure at the cost of resolution. High-order multi-feedback BP modulators do likewise, though their immunity to excess delay isn't as good as in their LP counterparts[2]. Finally, in fairness, even though the resolution of some of the ideal modulators in Figure 4.12 exceeds 16 bits, it is unlikely that GHz-speed modulators would achieve such a high resolution because other nonidealities such as thermal noise and clock jitter will almost surely limit performance more than quantization noise [Dia92a]. We will discuss the latter in the following chapter.

[2]The Matlab code written to do the transformations was unfortunately not sophisticated enough to handle BP modulators with non-coincident NTF zeros, though it seems reasonable to assume the results for such modulators would echo those seen in Figure 4.13.

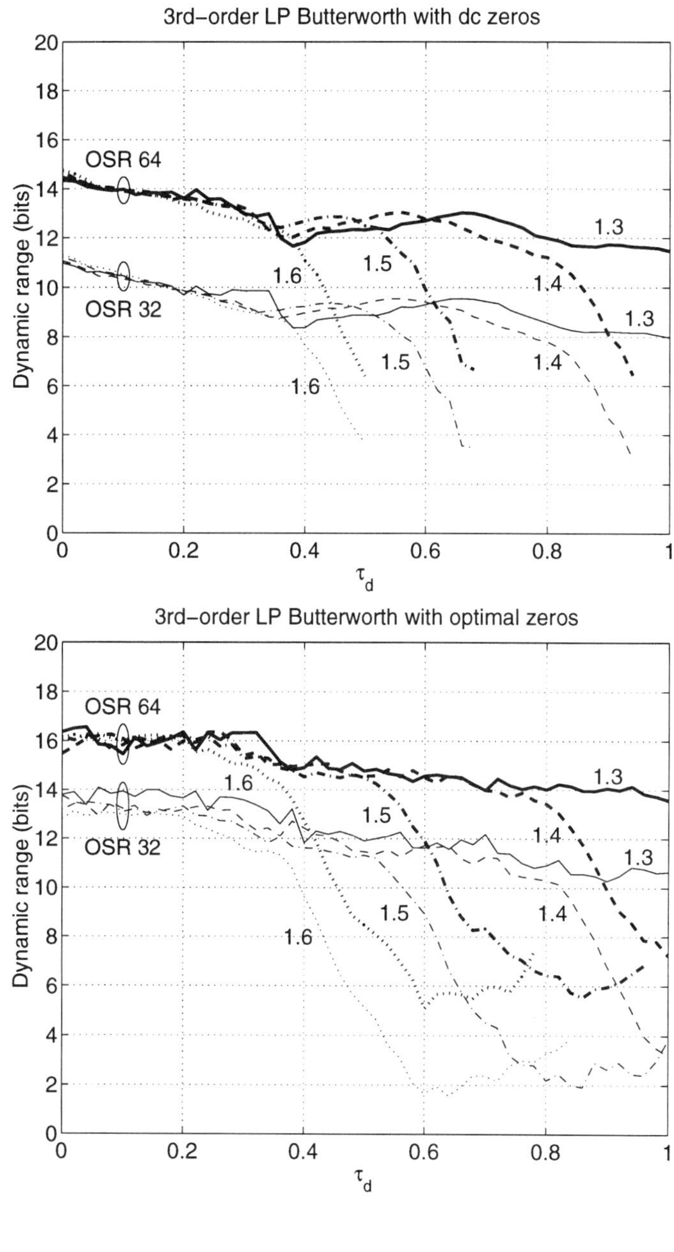

Figure 4.12. Dynamic range for high-order LP CT ΔΣMs. Numbers on curves are OOBG values.

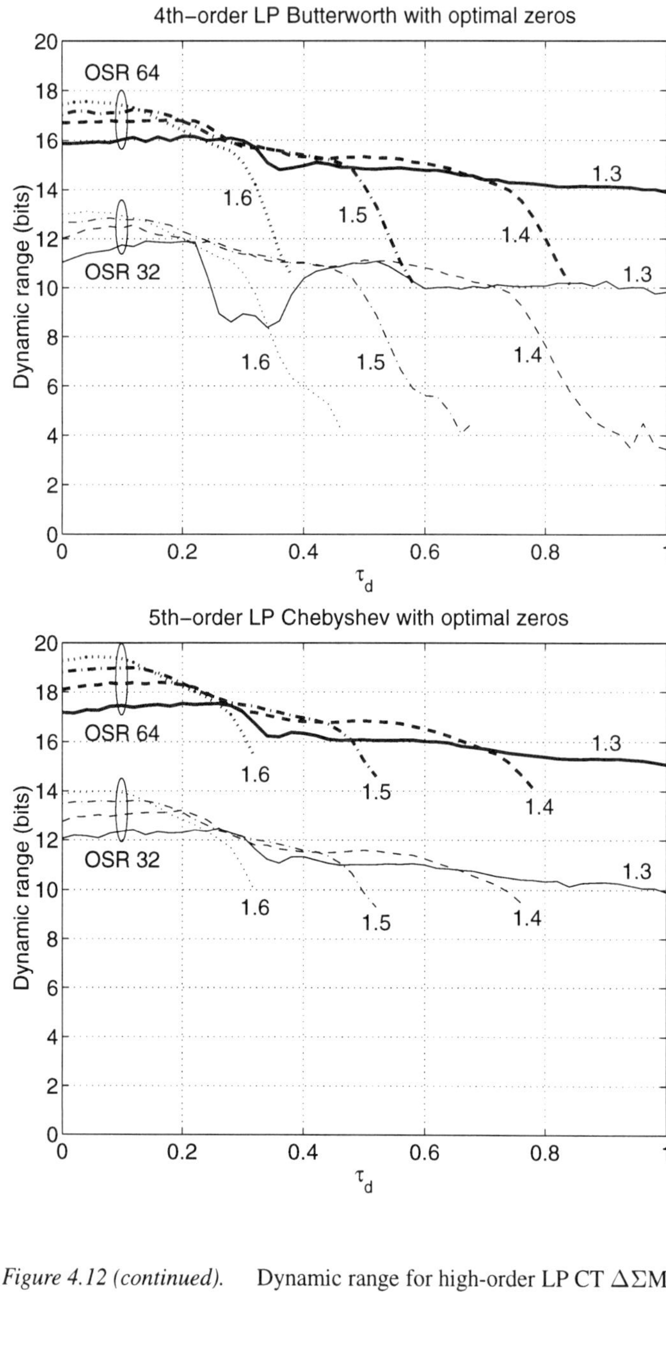

Figure 4.12 (continued). Dynamic range for high-order LP CT ΔΣMs.

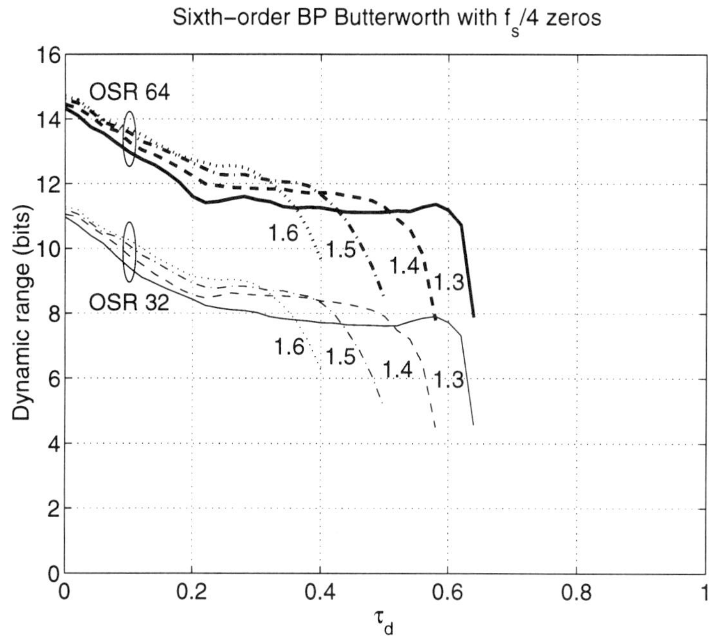

Figure 4.13. Dynamic range for sixth-order BP CT ΔΣM. Numbers on curves are OOBG values.

5. MODULATORS WITH A MULTIBIT QUANTIZER

Thus far we have concentrated on ΔΣMs employing a single-bit quantizer. It is known that multibit quantizers in DT designs improve stability, particularly for high-order designs [Nor97, Chap. 8]. If the previous section is any guide, we can hope that a multibit quantizer improves the immunity of CT designs to excess delay.

There is some improvement, but not a lot. Figure 4.14a shows the DR against excess delay for the second-order LP modulator for OSR $= 64$, while Figure 4.14b is for the fourth-order BP modulator. The thick lines are from Figure 4.8 and Figure 4.9a, the results for a 2-level (1-bit) quantizer, and the other lines are for 3-, 4-, 8-, and 16-level (1.5-, 2-, 3-, and 4-bit) quantizers. Generally, DR improves with quantizer resolution as expected, and furthermore the τ_d range over which the modulators remain stable improves a little with increasing quantizer resolution. Similar results are seen for the high-order LP modulators as for the second-order LP modulator. We see that the fourth-order BP circuit can be stable for τ_d close to 0.7 with a 4-bit quantizer compared to 0.5 for a 1-bit quantizer. Again, similar results are seen for the sixth-order BP modulator.

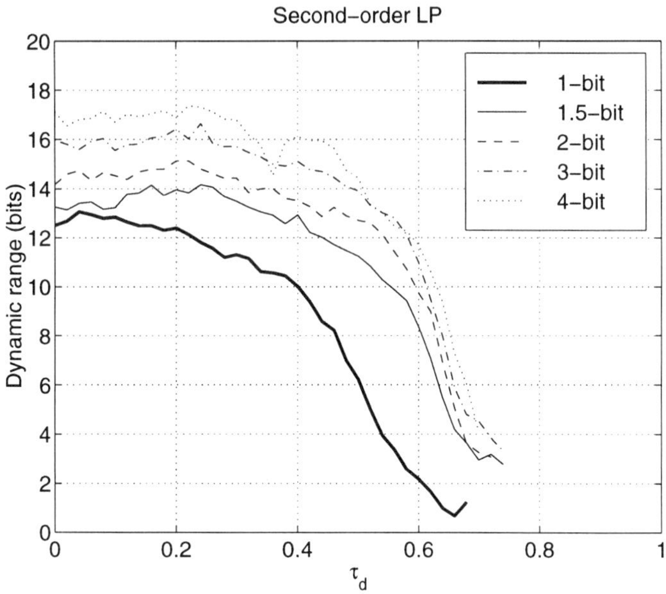

Figure 4.14a. Second-order LP modulator with a multibit quantizer.

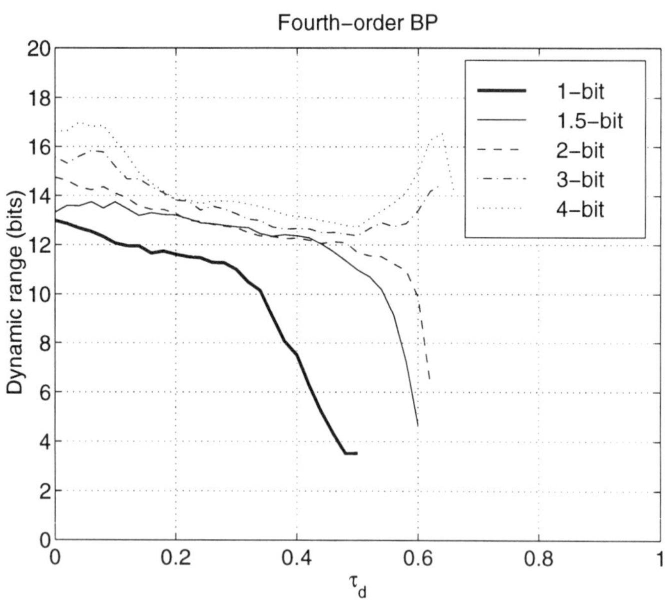

Figure 4.14b. Fourth-order BP modulator with a multibit quantizer.

We noted in Chapter 3 §1.2 the traditional problem in multibit designs: any level mismatches in the multibit feedback DAC are directly input-referred, thereby limiting the achievable performance. Implementing either DEM or digital post-correction on the same chip as a high clock-rate modulator might be a problem because both would require digital circuitry switching at f_s. This would cause a great deal of switching noise that might couple through the substrate into the forward modulator circuitry and degrade performance. Moreover, DEM would mean switching circuitry in the feedback path, which would add excess delay. Multibit quantizers are attractive both for stability and for reducing jitter sensitivity [Ada98], something we consider further next chapter. Nontheless, there have been no attempts to build a high-speed CT $\Delta\Sigma$M with a multibit quantizer, probably because of the difficulties just mentioned.

6. COMPENSATING FOR EXCESS LOOP DELAY

We have seen that it is possible to make a modulator immune to excess delay by choosing its OOBG appropriately. However, there exist methods of actually compensating for delay. We turn now to discussing them for single-bit designs, though the results are equally applicable to multibit designs. We explore some past proposals in more detail than previously reported, and also suggest some new methods.

6.1 DAC PULSE SELECTION

In §2., we considered the second-order LP $\Delta\Sigma$M with NRZ DAC pulses. A problem with this kind of pulse is that any excess loop delay $\tau_d > 0$ causes $\beta > 1$, which means the end of the pulse extends beyond $t = 1$. We saw in equations (4.7) through (4.11) that this increases the order of the resulting equivalent $H(z)$: in (4.11), $H(z)$ has the two poles at $z = 1$, but it acquires an additional pole at $z = 0$ for $\tau_d > 0$. Thus, the second-order modulator we tried to build actually has a *third* order loop filter[3]. In general, in any CT modulator with enough excess delay to push the falling DAC pulse edge past $t = 1$, the order of the equivalent DT loop filter is one higher than the order of the CT loop filter. Thus, a multi-feedback BP modulator using either an NRZ or HRZ pulse increases in order, as do the higher-order LP modulators from §4. with NRZ DACs. Another way to think about this increase in order is that it adds intersymbol interference: the DAC pulse from a previous symbol overlaps the current one.

[3]For small τ_d, the NTF has a pole and a zero close to one another which almost cancel, so the design appears approximately second-order in that case.

If we were to use DAC pulses with $\beta < 1$, then the pulses would extend past $t = 1$ only if the condition

$$\tau_d > 1 - \beta \tag{4.18}$$

held. This suggests the following for the second-order LP modulator in Figure 2.2: if we used an RZ DAC instead of an NRZ DAC, $H(z)$ would remain second-order for $\tau_d \leq 0.5$. If we knew exactly what τ_d is, we could select the feedback coefficients $\{k_2, k_1\}$ to get exactly the equivalent $H(z)$ from (2.7).

Example 4.5: For Figure 2.2, the loop filter is

$$\hat{H}(s) = \frac{k_2 + k_1 s}{s^2}. \tag{4.19}$$

Applying Table 4.1 to the partial fraction expansion of this for $(\alpha, \beta) = (\tau_d, \tau_d + 0.5)$ gives

$$H(z, \tau_d) = \frac{[4k_1 + k_2(3 - 4\tau_d)]z + [-4k_1 + k_2(1 + 4\tau_d)]}{8(z-1)^2}. \tag{4.20}$$

We wish for this to equal (2.7); equating powers of z in the numerator and solving yields

$$\{k_2, k_1\} = \{-2, -\frac{5}{2} - 2\tau_d\}. \tag{4.21}$$

Thus, for a given $\tau_d \leq 0.5$ and RZ DAC pulses, we can make our $\hat{H}(s)$ match exactly the desired $H(z)$ by *tuning* the parameter k_1. In the particular circuit of Figure 4.1, this is accomplished by changing the value of the current source in the rightmost differential pair DAC. □

It has long been recognized that it is sensible to use RZ DAC pulses in low pass CT $\Delta\Sigma$Ms [Ada86, Com91, Cha92, Nar94, Mit95]. Apart from the immunity to excess delay it afford us, an RZ DAC also alleviates intersymbol interference problems caused by asymmetric DAC pulse rise and fall times [Ada86], as we explained in Chapter 2 §3..

6.2 FEEDBACK COEFFICIENT TUNING

We just said that if there exists enough excess delay to push the falling edge of a DAC pulse past $t = 1$, the modulator order increases by one. Therefore, there will be $m + 1$ coefficients in the numerator of the equivalent $H(z)$; with only m feedback coefficients k, the system is not fully controllable via these ks alone. Previous examinations of loop delay in $f_s/4$ BP $\Delta\Sigma$Ms (notably [Sho96, §3.1.4] and [Gao97a]) have studied the system in Figure 2.4 using the modified $\mathcal{Z}$-transform and found the number of parameters in the numerator is m. It turns out that the multi-feedback architecture achieves a numerator coefficient of 0 for the z^{-1} term only because of perfect cancellation in the

$\tau_d = 0$ case. For $\tau_d \neq 0$ the cancellation is ruined so the coefficient of z^{-1} is nonzero, yet the modified $\mathcal{Z}$-transform incorrectly finds it to remain zero—yet another reason to avoid the modified $\mathcal{Z}$-transform here. There are actually $m+1$ rather than m numerator coefficients for $f_s/4$ BP modulators with excess delay.

Even though delay causing $\beta > 1$ means the system cannot be controlled perfectly with the ks, *some* degree of control can be exercised. We demonstrate the helpfulness of this on the fourth-order multi-feedback modulator in Figure 2.4.

> **Example 4.6**: Suppose there is a fixed excess delay of $\tau_d = 35\%$. Figure 4.9a shows that for OSR $= 64$, a DR of 9.9 bits is achieved using nominal k values, which can be found using a procedure like that in Example 2.5 to be
>
> $$\{k_{r4}, k_{r2}, k_{h4}, k_{h2}\} = \{-1.08678, -2.13388, 0.45016, 1.48744\} \quad (4.22)$$
>
> for RZ and HRZ DAC pulses. It is found that IBN $= -78$dB and MSA $= 0.34$ at this τ_d.
>
> Figure 4.15 shows how the performance of the modulator is affected when the ks are tuned one at a time away from their nominal values. By adopting a steepest-descent tuning approach where each k is tuned iteratively until the DR is maximized, we find that it is possible to improve the DR from 9.9 bits to 11.3 bits, still at $\tau_d = 0.35$. The IBN and MSA are both improved, IBN to -79dB and MSA to 0.74. The k values which give this performance are approximately
>
> $$\{k_{r4}, k_{r2}, k_{h4}, k_{h2}\} = \{-0.87, -1.83, 0.48, 1.89\}. \quad (4.23)$$
>
> The tuned k performance is still not as good as the 13 bits achieved at no excess delay in Figure 4.9a, but it is an improvement compared to the untuned k performance. □

An interesting thing happens when we tune the ks to maximize performance over a wide range of excess delay values.

> **Example 4.7**: Figure 4.16 compares the modulator DR for untuned k parameters from Figure 4.9a and tuned k parameters where the steepest-descent algorithm was applied for several different values of excess delay between 0 and 1. We see that it is possible to find k values which keep the modulator stable for the entire range of τ_d. What is perhaps more surprising is that performance worsens up to 50% excess delay, but then actually starts to improve again until there is a full sample delay, whereupon the performance becomes as good as it was for no delay at all! How can this apparently incongruous result be true?

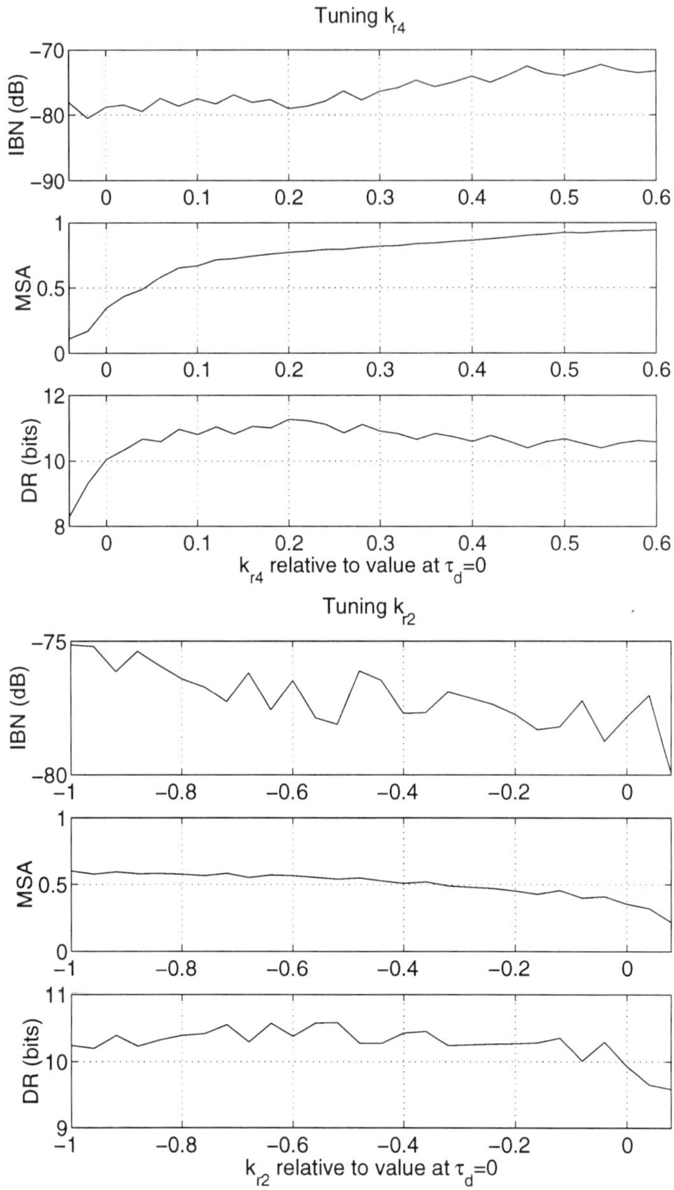

Figure 4.15. Fourth-order $f_s/4$ BP CT $\Delta\Sigma$M performance with 35% delay and feedback coefficient tuning.

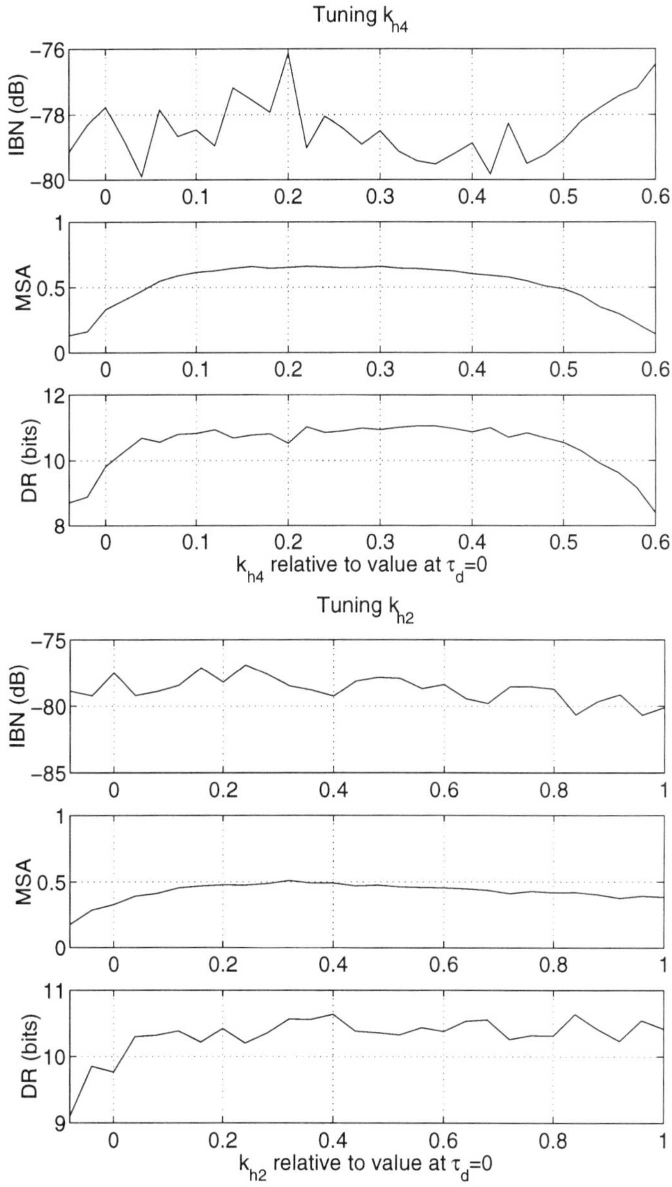

Figure 4.15 (continued). Fourth-order $f_s/4$ BP CT $\Delta\Sigma$M feedback tuning.

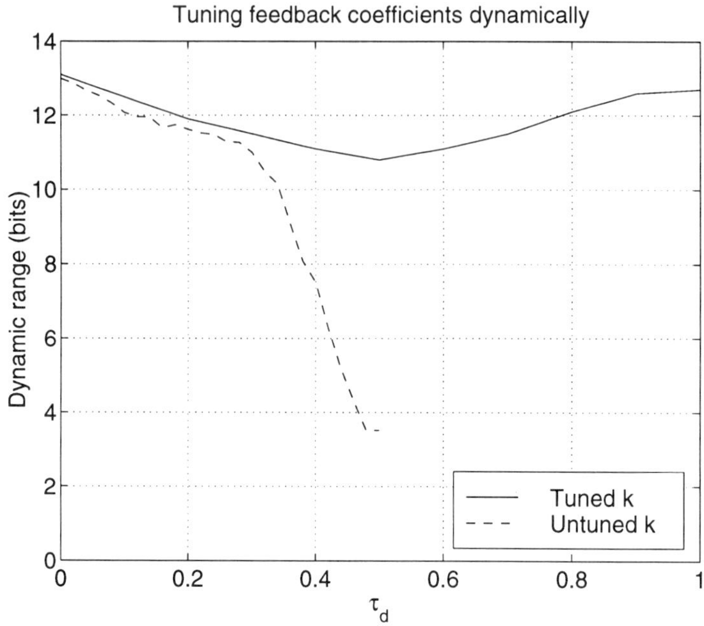

Figure 4.16. Multi-feedback BP modulator dynamic range with k tuning.

Recall $H_{BP}(z)$ in (2.15): the numerator was $2z^{-2} + z^{-4}$. The z^{-2} means there is a two-sample delay in the feedback; every $\Delta\Sigma$M must have at least one sample of delay in order to be causal. For NRZ DAC pulses, the equivalent $\hat{H}_{BP}(s)$ was (2.16):

$$\hat{H}_{BP}(s) = \frac{-1.0354s^3 + 1.0652s^2 - 1.3210s + 4.5661}{(s^2 + (\frac{\pi}{2})^2)^2}. \tag{4.24}$$

The two-sample delay is implicit in this equation. Note that we can write $H_{BP}(z)$ in an alternate manner:

$$H_{BP}(z) = \frac{2z^{-2} + z^{-4}}{(1 + z^{-2})^2} = z^{-1}\frac{2z^{-1} + z^{-3}}{(1 + z^{-2})^2}. \tag{4.25}$$

This suggests we could place a digital latch that provides one sample of delay (z^{-1}) prior to the DACs, and then find the equivalent $\hat{H}_{BP}(s)$ for the $H(z)$ with numerator $2z^{-1} + z^{-3}$.

What we're saying is, we have two choices for building a two-sample delay into the CT feedback loop: by doing $z \to s$ on an $H(z)$ with two delays in the numerator, *or* by providing a latch which adds one delay and doing $z \to s$ on an $H(z)$ with one delay in the numerator. These are denoted, respectively, the *zero* and *one digital delay* schemes in [Sho96]. This choice is peculiar to $f_s/4$ BP modulators; it does not exist for LP modulators or BP modulators at a different frequency because they invariably have a nonzero z^{-1} term in the numerator, and therefore $H(z)$ would become non-causal if we were to factor out a z^{-1} as we did in (4.25).

For each scheme, it is possible analytically to find the feedback ks which implement the desired $H(z)$:

$$\{k_{r4}, k_{r2}, k_{h4}, k_{h2}\} = \begin{cases} \{-1.0868, -2.1339, 0.4502, 1.4874\}, & \text{zero digital delay,} \\ \{-0.4502, -0.6339, 1.0868, 2.9874\}, & \text{one digital delay,} \end{cases} \tag{4.26}$$

where the first set of ks is from (4.22). The reason for the identical DR performance observed at both $\tau_d = 0$ and $\tau_d = 1$ is now clear: for $\tau_d = 1$, the optimal ks are those in the second row of (4.26), and the steepest-descent algorithm turns out to converge to values close to those. For $0 < \tau_d < 1$, the ks for optimal DR lie in between the zero and one digital delay values—compare, for example, (4.23) for $\tau_d = 0.35$ to (4.26)—though unfortunately the relationship between τ_d and the ks which optimize DR is not linear. For example, for $\tau_d = 0.5$, picking k values that lie exactly half way between the values in (4.26) leads to DR $= 9.2$ bits, though the steepest-descent algorithm found k values to make a modulator with DR $= 10.8$ bits. □

Figure 4.16 is strong encouragement to design the ks to be tunable, possibly even for on-line calibration against process and temperature variations. How to design a tuning algorithm to maximize DR that works on-chip, perhaps even while the modulator is operating, is an interesting topic for future research.

6.3 ADDITIONAL FEEDBACK PARAMETERS

If $\beta > 1$ causes the modulator order to increase from m to $m + 1$, and we only have m feedback coefficients, then it stands to reason that employing an additional feedback should restore full controllability to the system. This has been suggested in [Ben97]: in the block diagram of Figure 2.2, a third NRZ feedback was added whose output goes directly to a summing node after the second integrator (that is, immediately prior to the quantizer). To use this approach in a circuit architecture like Figure 4.1, where the quantizer input must be a voltage but summation is done with currents, we would have to add a transconductor followed by a current-to-voltage converter in between the second op amp and quantizer.

We can avoid adding components in the forward $\Delta\Sigma$M path by using an additional feedback with a *different* kind of DAC pulse. This is akin to the multiple feedbacks in the multi-feedback BP circuit, but it is believed that this has not been suggested previously for delay compensation in LP modulators.

Example 4.8: Consider again Figure 2.2: let us denote its NRZ feedback parameters $k_{n2} = k_2$ and $k_{n1} = k_1$, and let us suppose there is a third feedback which goes to the same summing node as k_{n1}: an HRZ DAC with coefficient k_{h1}. The z-domain equivalents for the NRZ pulses with excess delay have already been found in (4.9) and (4.10); to generalize them to feedbacks k_{n1} and k_{n2} instead of -1.5 and -1 is a trivial change to those equations. For an HRZ pulse delayed by τ_d, the z-domain equivalent is

$$\frac{k_{h1}}{s} \rightarrow \frac{k_{h1}(0.5-\tau_d)}{z-1} + z^{-1}\frac{k_{h1}\tau_d}{(z-1)}. \tag{4.27}$$

Combining this with (4.9) and (4.10) yields

$$H(z) = \frac{y_2 z^2 + y_1 z + y_0}{z(z-1)^2} \tag{4.28}$$

where $\{y_2, y_1, y_0\}$ are expressions involving $\{k_{n2}, k_{n1}, k_{h1}, \tau_d\}$. We wish for the numerator of this to equal $-2z^2 + z$ from (2.7), and Maple can be used to solve symbolically for the k values:

$$\begin{aligned} k_{n2} &= \frac{\tau_d^2+2}{\tau_d^2-2}, \\ k_{n1} &= \frac{-\tau_d^4+4\tau_d^3-12\tau_d^2+10\tau_d+4}{\tau_d^2-2}, \\ k_{h1} &= \frac{\tau_d^4-4\tau_d^3+13\tau_d^2-12\tau_d-2}{\tau_d^2-2}. \end{aligned} \tag{4.29}$$

Therefore, given the excess delay τ_d, we can get exactly the $H(z)$ in (2.7) by tuning the feedbacks to the values given in (4.29).

We could also use an HRZ pulse fed back to the first summer; this would give us different equations from (4.29), but still it would be possible to achieve the $H(z)$ in (2.7). However, we could *not* use an RZ pulse in place of an HRZ pulse. This is because for $\tau_d \leq 0.5$, the RZ pulse would not contribute to y_0 in (4.28): only k_{n2} and k_{n1} would, and thus to set $y_0 = 0$ (as (2.7) dictates) would require $k_{n2} = k_{n1} = 0$, which renders the feedback inoperational. □

How do we add an additional parameter to the BP multi-feedback architecture for delay compensation? Interestingly, adding an NRZ pulse to Figure 2.4 turns out not to work. This is because an NRZ pulse is a linear combination of RZ and HRZ pulses, so its feedback parameters are not independent. An independent pulse is needed—for example a pulse with $(\alpha, \beta) = (0.25, 1)$ in combination with any two of NRZ, RZ, and HRZ—but generating a pulse other than these latter three might be nontrivial at high speed.

7. SUMMARY

Excess loop delay in a CT ΔΣM is a delay between the sampling clock edge and the change in output bit as seen at the feedback point in the modulator. It arises because of the nonzero switching time of the transistors in the feedback path, and is significant because it alters the equivalence between the CT and DT representations of the loop filter, $\hat{H}(s)$ and $H(z)$. Its effect on performance is noticeable if the sampling clock speed is an appreciable fraction (10% or more) of the maximum transistor switching speed; this is becoming more likely nowadays as desired conversion bandwidths increase and delta–sigma modulation with an aggressively-high clock rate relative to the transistor switching speed is considered for the converter architecture.

If excess delay is not designed for, then as excess delay increases as a fraction of the clock period, second-order LP and fourth-order $f_s/4$ BP modulators will suffer in terms of in-band noise, maximum stable input amplitude, and dynamic range. Higher-order LP designs seem more robust if designed using NTF prototyping because there is a parameter, the out-of-band gain, which can be selected to give some immunity to excess delay. Higher-order BP designs are also more robust than lower-order ones, but a multi-feedback $f_s/4$ BP design is always found to be less immune to excess delay than the corresponding LP design. The use of a multibit quantizer is somewhat helpful, though incorporating the usually-needed correction circuitry for a feedback DAC with mismatched levels is nontrivial for high-speed designs.

It is imperative to recognize the presence of excess delay in a high-speed design and take it into account in the design process. Choosing the right DAC pulse shape in combination with tuning of the feedback parameters (either in the design phase or automatically on-line) can greatly mitigate the performance loss due to delay. In fact, taking excess delay into account renders it effectively a nonproblem. Our study has contributed a number of useful new results: we use a transformation method which treats the delay as occurring after the quantizer, not before as the modified $\mathcal{Z}$-transform does; we consider both the change in noise *and* MSA with delay; we show how delay affects high-order designs, as well as multibit designs; and we demonstrate compensation methods based on RZ DAC pulses and additional feedbacks using independent DAC pulses.

Chapter 5

CLOCK JITTER AND QUANTIZER METASTABILITY

We learned in Chapter 2 §3. and §4.2 that any nonuniformity in the DAC pulses results in degraded CT modulator performance. In this chapter we study two major effects that contribute to this degradation: timing jitter in the quantizer clock, or clock jitter, and quantizer metastability.

1. PRELIMINARIES

In this section we explain why CT $\Delta\Sigma$Ms are particularly susceptible to jitter errors, and we explain a little bit about the simulation program we used to generate the results in this chapter.

1.1 JITTER IN CT MODULATORS

Suppose there is quantizer clock jitter. On the left of Figure 5.1, a typical

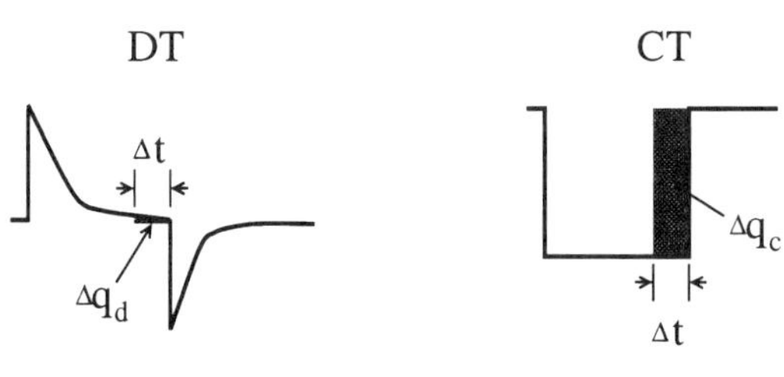

Figure 5.1. Clock jitter effect in DT vs. CT DAC currents.

feedback current waveform for an SC DT $\Delta\Sigma$M is depicted. Most of the charge transfer occurs at the start of the clock period, so that the amount of charge Δq_d lost due to a timing error is relatively small. By contrast, the right of Figure 5.1 shows the DAC output current in a CT circuit such as Figure 4.1; there, charge is transferred at a constant rate over a clock period, and so charge loss Δq_c from the same timing error is a larger proportion of the total charge. Moreover, in a DT design, jitter in the input sample-and-hold (S/H) clock means only the input waveform is affected. In a CT design, the sampling occurs at the quantizer rather than the input, which means the jitter affects the sum of the input plus quantization noise—a signal with considerably more power than the input alone. Hence, CT $\Delta\Sigma$Ms are more sensitive to clock jitter than DT designs [vdZ96].

Clock jitter causes a slight random variation in the amount of charge fed back per clock cycle. Put another way, it is akin to adding a random phase modulation to the output bit stream. In an oversampled converter, the spectrum of the output stream is very noisy outside the (narrow) signal band; a random phase modulation causes the noise outside the signal band to fold into the signal band, raising the converter noise floor and degrading its resolution. In this chapter we quantify this degradation for a fully-integrated high-speed modulator, that is, one that is clocked by a typical on-chip VCO with a given a phase noise specification.

Even if the sampling were clock perfectly uniform, it is nonetheless possible for there to exist a variation in the feedback charge. This happens because a real quantizer contains a regenerative circuit with a finite regeneration gain. Therefore, quantizer inputs with a magnitude near zero will take longer to resolve than inputs with a large magnitude—this is the classic problem of metastability in digital latches. We illustrated this in Chapter 2 §4.2 for a real GHz-speed modulator.

In a $\Delta\Sigma$M, the input to the quantizer is decorrelated from the modulator input to the degree that it appears random; hence, the times when the quantizer input is near zero also appear random. This means that at certain unpredictable sampling instants, slightly more charge is transferred for the previous clock period and slightly less for the next period. Again, the effect is to modulate out-of-band noise into the signal band and degrade converter resolution. Here, we expand on our previous work in this area [Che97] by generalizing it to different orders of modulator and proposing methods to overcome the effects of quantizer metastability[Che99a].

1.2 MODULATOR ARCHITECTURE

A block diagram of the architecture of the modulators we simulate in this chapter is shown in Figure 5.2. It is similar to, but more generalized than, Figure 4.10. Each of the gain blocks $G_i(s)$ will typically be the same except possibly for the gains; thus, for a low pass modulator, the blocks can be represented with continuous-time integrators

$$G_i(s) = \frac{A_i}{sT_s} \tag{5.1}$$

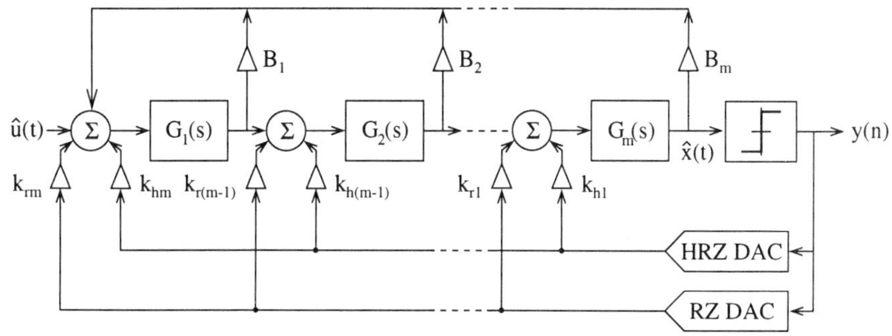

Figure 5.2. Block diagram for general CT ΔΣM architecture.

and for a band pass modulator, they will be resonators

$$G_i(s) = \frac{A_i s T_s}{s^2 T_s^2 + \omega_i^2}. \tag{5.2}$$

Without loss of generality, we will sometimes assume the sampling frequency is $f_s = 1\text{Hz}$, which simplifies the notation in (5.1) and (5.2) by making $T_s = 1$. Modulator behavior is unchanged so long as the proper scaling is applied to all circuit parameters.

The quantizer drives two separate DACs of the RZ and HRZ varieties whose levels at feedback points are set by coefficients $\{k_{rm}, k_{r(m-1)}, \ldots, k_{r1}\}$ and $\{k_{hm}, k_{h(m-1)}, \ldots, k_{h1}\}$, respectively. A modulator with an NRZ DAC can be effected by setting $k_{ri} = k_{hi}$, while one with an RZ DAC only would have all $k_{hi} = 0$. We have both types of DAC in Figure 5.2 separately tunable for when we wish to implement multi-feedback BP modulators.

1.3 SIMULATION METHOD

The time-domain state equations are coded in a C program and integrated numerically using a variable time step fourth-order Runge-Kutta (RK4) method [Pre92]. For example, for an LP modulator with gain blocks given by (5.1), the state equations are

$$\frac{1}{A_i}\frac{dx_i}{dt} = \begin{cases} B_1 x_1 + \cdots + B_m x_m + k_m y + u, & i = 1 \\ x_{i-1} + k_{m+1-i} y, & i = 2, \ldots, m \end{cases} \tag{5.3}$$

where $k_i = k_{ri}$ during the first half of a period and $k_i = k_{hi}$ during the second half. At every clocking instant, the quantizer output is evaluated; the power spectrum of N output bits is calculated by the program using the periodogram of the FFT, and periodograms from any number of runs with random initial states may be averaged to yield a fairly smooth spectrum from which the SNR may be found.

Moreover, certain nonidealities of interest in this chapter are implemented. In particular:

- the sampling instant can be affected by jitter;
- the DAC pulses can have excess delay and nonzero rise time;
- the quantizer can exhibit hysteresis and/or metastability.

Other nonidealities such as integrators with finite dc gains (or resonators with finite Qs) and finite gain block output swing can also be modeled without much difficulty.

The principal advantage of using C is that it runs very fast compared to, for example, block-diagram level simulation in a circuit simulator like SPICE, though it is slower and more error-prone to code. Presently we show that we can get acceptable agreement between the C program and a transistor-level simulation in SPICE with several orders of magnitude increase in simulation speed. The effort spent on the coding will thus appear to be justified.

2. WHITE CLOCK JITTER

Suppose the sampling times for N output bits are given by

$$t_n = nT_s + \beta_n,\ n = 0, 1, \ldots, N-1, \tag{5.4}$$

and for the moment, let the β_n be i.i.d. random variables with variance σ_β^2. Let us quantify the reduction in SNR through out-of-band noise modulation into the signal band for various modulators.

2.1 LP MODULATORS WITH NRZ FEEDBACK

Figure 5.2 can simulate low pass modulators with NRZ feedback by making the gain blocks integrators as in (5.1) and setting $k_{ri} = k_{hi} = k_i$. If all the integrators have gain $A_i = 1$, the loop filter implemented by the circuit for $m \geq 2$ is (4.17):

$$\hat{H}(s) = \frac{\sum_{i=1}^{m}(\frac{1}{s})^i[k_i - \sum_{j=1}^{i-1} k_{i-j}B_j]}{1 - \sum_{i=1}^{m} B_i(\frac{1}{s})^i}. \tag{5.5}$$

We can choose a loop filter $H(z)$ in the DT domain using any method we please and use the impulse-invariant transform to find the equivalent $\hat{H}(s)$ as we did in Chapter 2. It is then a trivial matter to pick k_i and B_i in (5.5).

Example 5.1: We have already said that the standard double integration modulator has a DT loop transfer function $H(z)$ and equivalent CT $\hat{H}(s)$ for NRZ DAC pulses given by

$$H(z) = \frac{-2z+1}{(z-1)^2} \leftrightarrow \hat{H}(s) = \frac{-1.5s-1}{s^2} \tag{5.6}$$

where $T_s = 1$ for simplicity. For integrators with gains $\{A_1, A_2\} = \{1, 1\}$,

$$\{k_2, k_1\} = \{-1, -1.5\},\ \{B_1, B_2\} = \{0, 0\} \tag{5.7}$$

are found from (5.5). This modulator was depicted in Figure 2.2. □

Example 5.2: A third-order modulator designed using NTF prototyping where NTF(z) has Butterworth poles, an out-of-band gain of 1.5, and zeros spread to minimize quantization noise in the signal band assuming OSR $= 32$ can be found to have $H(z)$ and $\hat{H}(s)$ for NRZ DAC pulses given by

$$\begin{aligned} H(z) &= \frac{-0.7874z^2 + 1.3085z - 0.5569}{z^3 - 2.9942z^2 + 2.9942z - 1} \\ \leftrightarrow \quad \hat{H}(s) &= \frac{-0.6702s^2 - 0.2407s - 0.0458}{s^3 + 0.0058s}. \end{aligned} \tag{5.8}$$

Choosing all integrator gains to be 1 yields

$$\begin{aligned} \{k_1, k_2, k_3\} &= \{-0.0831, -0.5021, -1.4659\}, \\ \{B_1, B_2, B_3\} &= \{0, -5.7830 \times 10^{-3}, 0\}. \end{aligned} \tag{5.9}$$

If we were interested in actually building this, we might find the range of these values (from 0.5×10^{-3} up to 1.5) too wide to be practical. Choosing the integrator gains as $\{A_1, A_2, A_3\} = \{0.2, 0.2, 1\}$ instead gives a smaller spread in the resulting k_i and B_i. Rewriting (5.5) for $A_i \neq 1$ and solving gives

$$\begin{aligned} \{k_1, k_2, k_3\} &= \{-1.0479, -1.2033, -0.6702\}, \\ \{B_1, B_2, B_3\} &= \{0, -0.1446, 0\}. \end{aligned} \tag{5.10}$$

Of course, for the purposes of simulation, either will work. □

Performing the same calculation for fourth-order Butterworth pole and fifth-order Chebyshev pole NTFs both with out-of-band gain 1.5 and optimally-spread zeros, and simulating those systems with different values of jitter standard deviation σ_β, results in the output spectra shown in Figure 5.3. These are 256 averaged 8192-point Hann-windowed periodograms whose x axes span a frequency range from 0 to $f_s/32$. The input tone was -20dB relative to full scale in bin 19 ($2.32 \times 10^{-3} f_s$). The β_n were normally-distributed. We indeed see the deep notch in the quantization noise gradually filled in with white noise with a power proportional to $10\log_{10}(\sigma_\beta/T_s)^2$.

The problem of white clock jitter in LP NRZ modulators has been studied in the past [Ris94, Dia92a]; we review the results briefly here to explain our

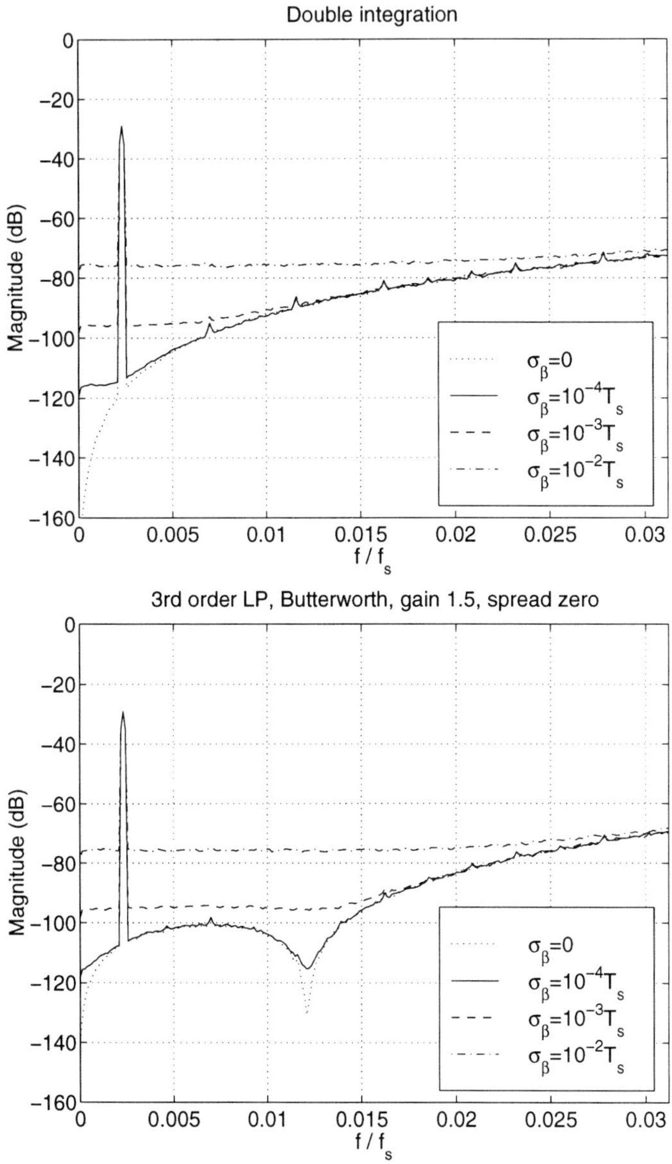

Figure 5.3. Output spectra for NRZ LP modulators with clock jitter.

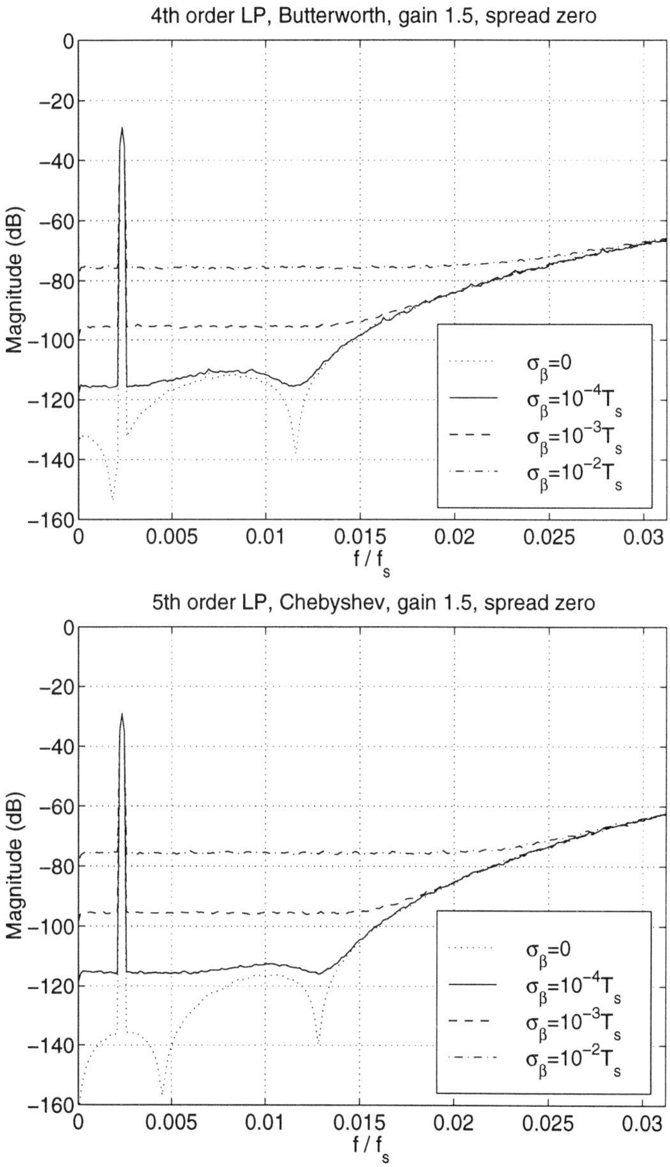

Figure 5.3 (continued). Output spectra for NRZ LP modulators with clock jitter.

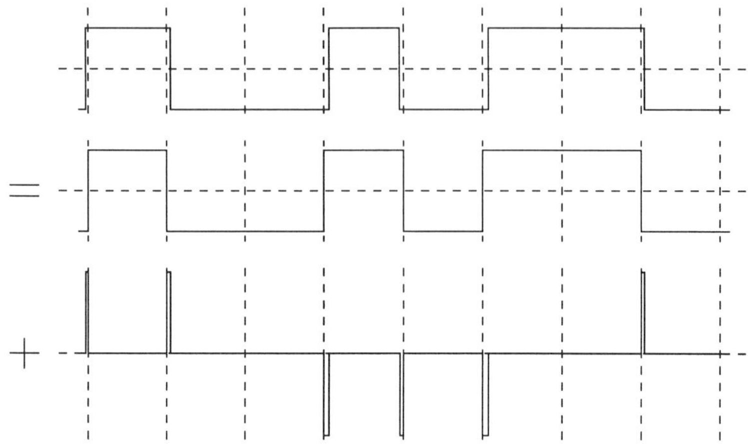

Figure 5.4. Equivalent representations of a jittered bit stream.

observations. The effect of jitter in an NRZ modulator can be explained by considering Figure 5.4. The output bit stream with jitter shown in the top diagram is equivalent to the sum of an unjittered bit stream (the middle diagram) and a stream of pulses, which we call the *error sequence*, resulting from the jitter (the bottom diagram). Because the Fourier transform is a linear operation, the output spectrum for the top signal must equal the sum of the spectra of the bottom two signals. The error sequence can be represented by [Ris94]

$$e_{NRZ}(n) = [y(n) - y(n-1)]\frac{\beta_n}{T_s} \tag{5.11}$$

where $y(n)$ is the nth output bit. For wideband uncorrelated jitter, this error will be almost white, in which case we may write

$$\sigma_e^2 \approx \sigma_{\delta y}^2 \times \frac{\sigma_\beta^2}{T_s^2}. \tag{5.12}$$

In other words, the variance of the error sequence is the product of the variance of $\delta y \equiv y(n) - y(n-1)$ and the jitter variance relative to the clock period. For N output bits, we expect the noise per periodogram bin to be

$$10\log_{10}\left(\frac{2\sigma_{\delta y}^2 \cdot 2\sigma_\beta^2}{NT_s^2}\right) - 7.27 \text{ dB} \tag{5.13}$$

where the factors of 2 in the numerator arise because we are taking the one-sided power spectrum and where 7.27 is the sum of $10\log_{10} 2 = 3.01$ dB (the power spectrum is rms power) and $10\log_{10} 0.375 = 4.26$ dB (0.375 is the gain

Table 5.1. Simulated and calculated LP NRZ modulator performance for $\sigma_\beta = 10^{-2}T_s$ in Figure 5.3.

Modulator	$\sigma_{\delta y}$	Baseband noise per bin		SNR for OSR = 32	
		Simulated	(5.13)	Simulated	(5.14)
Double integration	1.674	−75.8dB	−75.9dB	27.5dB	27.6dB
3rd order Butterworth	1.750	−75.6dB	−75.5dB	27.3dB	27.2dB
4th order Butterworth	1.739	−75.5dB	−75.6dB	27.4dB	27.2dB
5th order Chebyshev	1.731	−75.5dB	−75.6dB	27.5dB	27.3dB

of a Hann window). Moreover, if the SNR in baseband is completely limited by white jitter noise rather than noise-shaped quantization noise, then for an input amplitude V_{in} we can write [Ris94]

$$\text{SNR}_{NRZ} = 10\log_{10}\frac{\text{OSR}\cdot V_{in}^2/2}{\sigma_{\delta y}^2\left(\frac{\sigma_\beta}{T_s}\right)^2}\ \text{dB}. \tag{5.14}$$

Table 5.1 shows the agreement between calculated and simulated values of (5.13) and (5.14) for the four modulators in Figure 5.3. The theory for low pass NRZ modulators is confirmed.

2.2 MODULATORS WITH RZ AND/OR HRZ FEEDBACK

We saw last chapter that RZ DACs are beneficial for reducing intersymbol interference and excess delay problems in LP modulators. As well, multi-feedback BP modulators use both RZ and HRZ DACs in the same circuit. How does jitter affect the spectrum of a modulator using RZ (and possibly HRZ) DACs? Let us first choose circuit coefficients for typical modulators.

Example 5.3: The CT loop filter $\hat{H}(s)$ for a double integration LP modulator with RZ DAC pulses was found in Example 2.2 to be

$$H(z) = \frac{-2z+1}{(z-1)^2} \leftrightarrow \hat{H}(s) = \frac{-2.5s-2}{s^2}. \tag{5.15}$$

Picking the integrator gains to be 1 and the B_is to be zero leaves

$$\{k_2, k_1\} = \{-2, -2.5\}, \tag{5.16}$$

which we found in (4.21) for $\tau_d = 0$. □

Example 5.4: We considered the design of a fourth-order $f_s/4$ BP modulator in Example 4.7 already. The feedback coefficients for resonators in (4.22) were

$$\{k_{r2}, k_{r1}, k_{h2}, k_{h1}\} = \{-1.0868, -2.1339, 0.4502, 1.4874\}, \tag{5.17}$$

and these are correct for resonators with gains $\{A_1, A_2\} = \{\pi/2, \pi/2\}$ and $\omega_0 = \pi/2$ in (5.2). □

Output spectra for simulations of each system are shown in Figures 5.5a and 5.5b. Once again, these are 256 averaged 8192-point Hann-windowed periodograms. For a jitter standard deviation $\sigma_\beta = 10^{-2}T_s$, we found a baseband noise of -75.8dB per bin in the double integration NRZ modulator, while for the double integration RZ modulator in Figure 5.5a the value is -71.3dB, and for the BP modulator in Figure 5.5b the value is -66.7dB. Where do the new values come from?

Figure 5.6 shows the same bit sequence $\{+1, +1, -1, +1, -1\}$ as output by the same modulator with three different DACs: NRZ, RZ, and a combination of both RZ and HRZ DACs. The solid rectangles show edges which are affected by jitter. We may distinguish the three cases as follows.

- In an NRZ modulator, jitter only matters when the output changes sign—the error sequence $e_{NRZ}(n)$ is nonzero only at those times, c.f. (5.11). The energy in the error sequence is proportional to $\delta y^2 = [y(n) - y(n-1)]^2 = 4$ for a modulator with ± 1 outputs. For the double integration NRZ modulator in Table 5.1 we found $\sigma_{\delta y} = 1.673$, and the formula for variance is

$$\sigma_{\delta y}^2 = \frac{\sum \delta y^2 - \frac{(\sum \delta y)^2}{N}}{N-1} \approx \frac{4N_{\delta y}}{N} \tag{5.18}$$

  for large N where $N_{\delta y}$ is the actual number of output bit transitions. We have $\sigma_{\delta y}^2 = 2.80$ and can estimate $N_{\delta y}/N = \sigma_{\delta y}^2/4 = 0.70$ for that modulator.

- In an RZ modulator, both the rising and the falling edge of the pulse occur *every* clock cycle, so jitter affects a total of $2N$ edges. The energy per edge is $[\pm 1 - 0]^2 = 1$, one quarter as much as in the NRZ case. But now, energy is being transferred over only half a clock cycle; σ_β is therefore twice as large relative to the energy transfer period in an RZ modulator.

- In a modulator employing RZ and HRZ pulses of opposite sign, as is the case in a multi-feedback BP modulator, there are now N edges at half clock cycles when going from the RZ to the HRZ pulse, and edges at half cycles where the output bits $y(n-1)$ and $y(n)$ are the same. These edges have energy 4 as in the NRZ case, and σ_β is twice as large relative to the energy transfer period as in the RZ case. In simulation, we find $\sigma_{\delta y} = 1.405$ for the BP modulator, so that $N_{\delta y}/N = 0.494$ from (5.18). The total number of edges is then $N + N(1 - N_{\delta y}/N) = 1.506$.

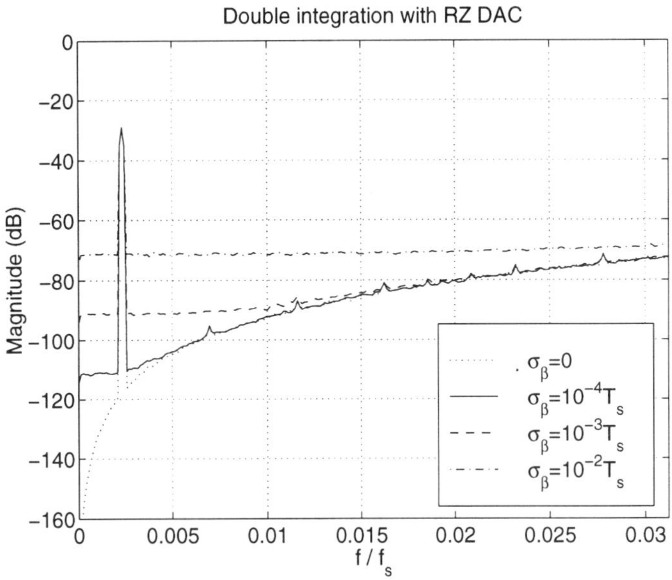

Figure 5.5a. Output spectra for LP RZ modulator.

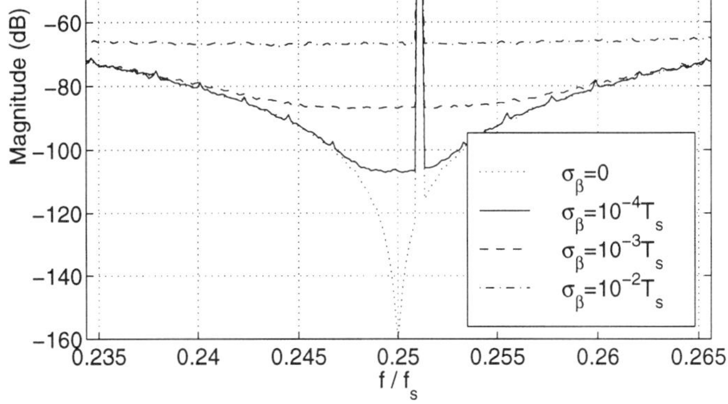

Figure 5.5b. Output spectra for BP RZ/HRZ modulator.

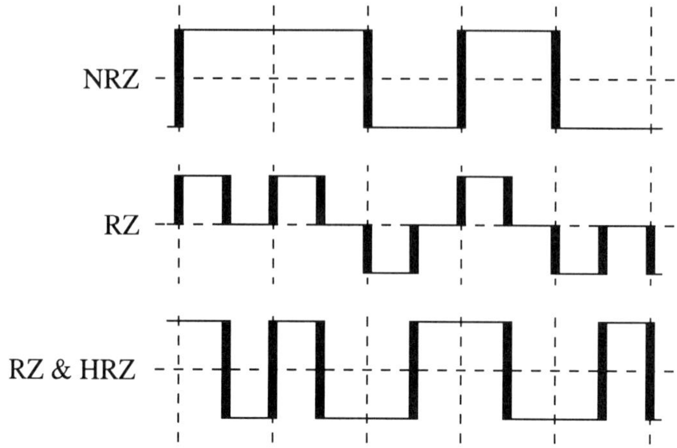

Figure 5.6. Error sequence energy in different types of modulator.

Taking all this into account, we may estimate an effective value of $\sigma^2_{\delta y}$ in (5.12):

$$\sigma^2_{\delta y} = \begin{cases} \frac{0.70N\times 4}{N} & = \quad 2.80, \quad \text{NRZ} \\ \frac{2N\times 1}{N}\times 2^2 & = \quad 8.00, \quad \text{RZ} \\ \frac{1.506N\times 4}{N}\times 2^2 & = \quad 24.10, \quad \text{RZ \& HRZ.} \end{cases} \tag{5.19}$$

Therefore, we expect the RZ LP modulator to be $10\log_{10}(8.00/2.80) = 4.6\text{dB}$ worse than the NRZ LP modulator and the BP modulator to be 9.3dB worse than the NRZ LP modulator. This is very close to what we observed (4.5dB and 9.2dB) in Figures 5.5a and 5.5b.

Generally, clock jitter affects RZ and/or HRZ modulators more severely than modulators employing just NRZ feedback. A good rule of thumb is, jitter noise will be 6 dB (1 bit) worse in the band of interest.

3. CLOCKING WITH A VCO

Although it is interesting theoretically, the analysis in §2. is not terribly relevant in practice. The problem is, if we are trying to use a high-speed CT $\Delta\Sigma$M in a practical circuit it will likely be clocked on-chip with an integrated VCO. Sampling instants as given in (5.4) are not what a real VCO provides—the jitter instants β_n from a VCO are not well-modeled as i.i.d. random variables. Figure 5.7a shows 256 averaged 8192-point unwindowed periodograms of a sine wave carrier sampled by an ideal S/H four times per period (i.e., $f_c = f_s/4$) with a jittered clock given by (5.4) and $\sigma_\beta = 10^{-4}T_s$. That kind of jitter, which we will denote *independent jitter*, adds white noise skirts to the carrier. A VCO produces skirts that are nonwhite.

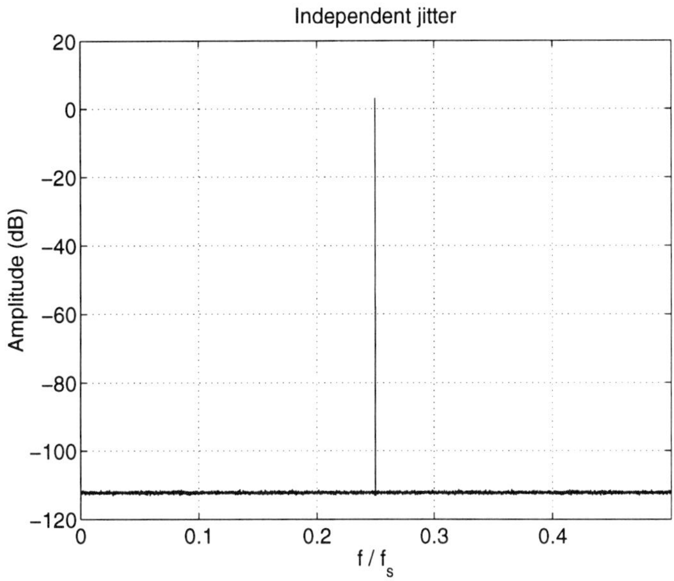

Figure 5.7a. Unwindowed spectrum of sine wave carrier sampled with independent jitter.

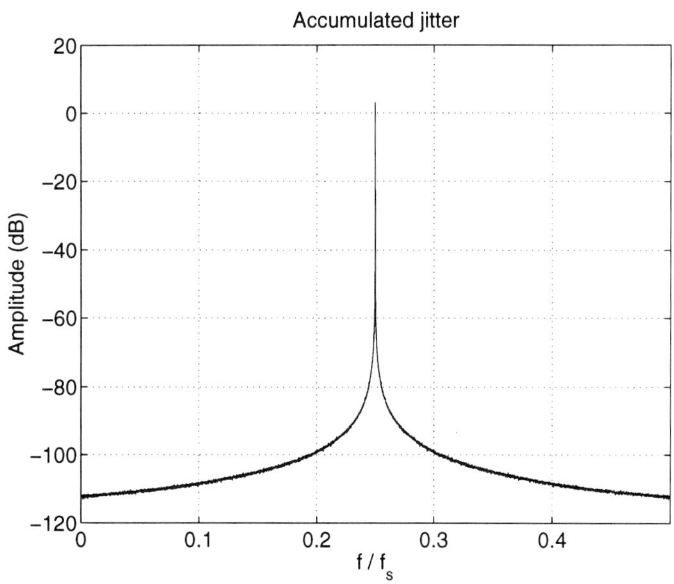

Figure 5.7b. Unwindowed spectrum of sine wave carrier sampled with accumulated jitter.

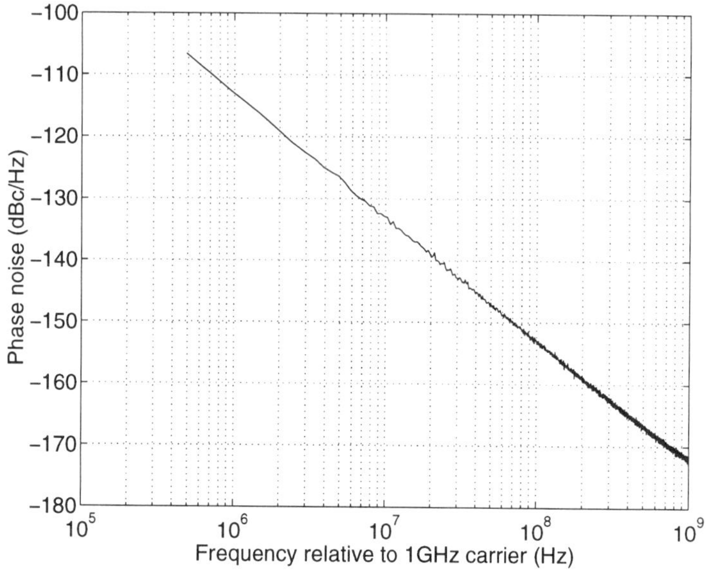

Figure 5.8. Phase noise plot for accumulated jitter sampling of sine wave.

3.1 MODELING VCO PHASE NOISE

We can modify (5.4) to produce nonwhite skirts fairly easily using a result due to Berkovitz and Rusnak [Ber92]. Suppose the sampling instants are instead given by

$$t_n = nT_s + \sum_{i=0}^{n} \beta_i, \; n = 0, 1, \ldots, N-1 \tag{5.20}$$

where β_i are still i.i.d. We denote this *accumulated jitter* because it contains a running sum, and a sine wave sampled four times per period with a jittered clock given by (5.20) has the spectrum shown in Figure 5.7b. Plotting the magnitude of the skirts relative to the carrier with a logarithmic frequency scale, as is customarily done in a VCO phase noise plot, yields the graph in Figure 5.8, where we have assumed the sine wave has a frequency of $f_c = 1\text{GHz}$. The sideband power has a $1/f^2$ dependence—as is exactly the case in an integrated VCO [Haj98]. A VCO also has a $1/f^3$ region close to the carrier, and a white noise floor far from the carrier, but (5.20) at least gives a reasonable approximation of a VCO phase noise over frequencies an intermediate distance from the carrier. Phase noise in a VCO is usually specified as n_c dBc/Hz at an offset f_n from the carrier f_c. Fortuitously, this f_n is usually in the $1/f^2$ region of the phase noise.

A typical achievable value of n_c is [Lee66, Dau98]

$$n_c = -100 + 20\log_{10} f_c \text{ dBc/Hz at 100kHz offset} \tag{5.21}$$

for f_c in GHz. How can we relate this to σ_β? We are interested in the case of $f_c = f_s$, since for clocking a $\Delta\Sigma$M, all that matters is the jitter of the zero crossings of the carrier, yet we are constrained mathematically in a periodogram to $f_c < f_s/2$. However, it is found that altering the ratio f_c/f_s moves the phase noise in Figure 5.8 proportionally to $10\log_{10} f_c/f_s$, so we can use this to extrapolate to $f_c = f_s$. After some experimentation with normally-distributed β_n, it is found that using

$$\frac{\sigma_\beta^2}{T_s^2} \approx \frac{f_n^2 \times 10^{n_c/10}}{2f_c} \tag{5.22}$$

gives a phase noise of n_c relative to the carrier at f_n offset from f_c.

Therefore, we can simulate the effect of clocking a $\Delta\Sigma$M with a VCO meeting a certain phase noise spec by using sampling instants with accumulated jitter (5.20) and a variance given by (5.22).

3.2 EFFECT OF ACCUMULATED JITTER ON PERFORMANCE

Example output spectra for four different types of modulator are depicted in Figure 5.9. These are the third-order Butterworth NRZ, fifth-order Chebyshev NRZ, double integration RZ, and fourth-order $f_s/4$ multi-feedback BP $\Delta\Sigma$Ms from Figures 5.5a and 5.5b. All simulations used $\sigma_\beta = 10^{-2}T_s$, and for contrast both independent *and* accumulated jitter spectra are plotted along with unjittered spectra.

There are two traits in the accumulated jitter spectra worth noting. First, accumulated jitter whitens the in-band spectrum in much the same way as independent jitter—this is not unexpected because *any* clock spectral impurities will randomly modulate out-of-band noise into the signal band. The white noise floor seems to be between about 1–5dB lower for accumulated jitter compared to independent with the same σ_β; for a given modulator, simulations show this number to be roughly constant for any values of σ_β, input frequency, and input amplitude. Second, the dash-dot lines on each graph show the spectrum of a sine wave with the same frequency as the input tone that has been sampled by a S/H circuit clocked with the same clock as the quantizer. The skirts on the tone appear directly in the output spectrum so long as they are higher than the white noise floor. This, too, is logical.

Note the significance of the observation in §3.1 that the height of the skirts is proportional to $10\log_{10} f_c/f_s$: as the input tone moves to higher frequencies, the skirts become higher relative to the tone. Thus, an LP modulator with a

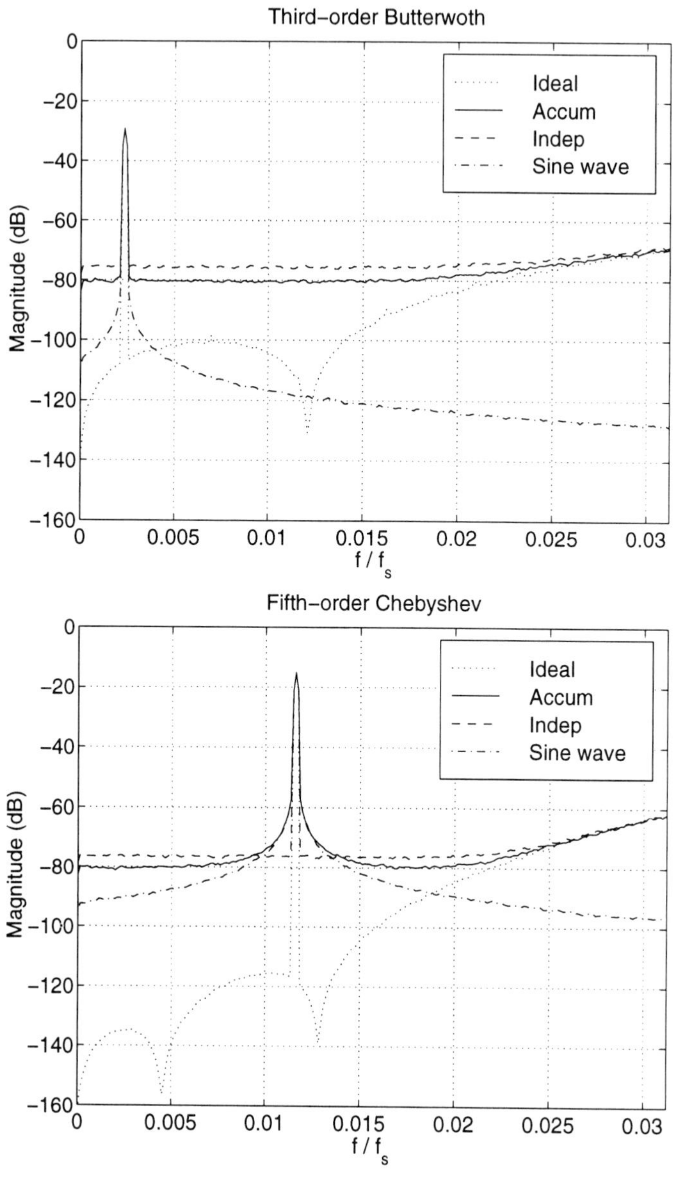

Figure 5.9. Comparison of $\Delta\Sigma$M spectra for independent and accumulated jitter, including spectrum of single tone at output of jittered S/H.

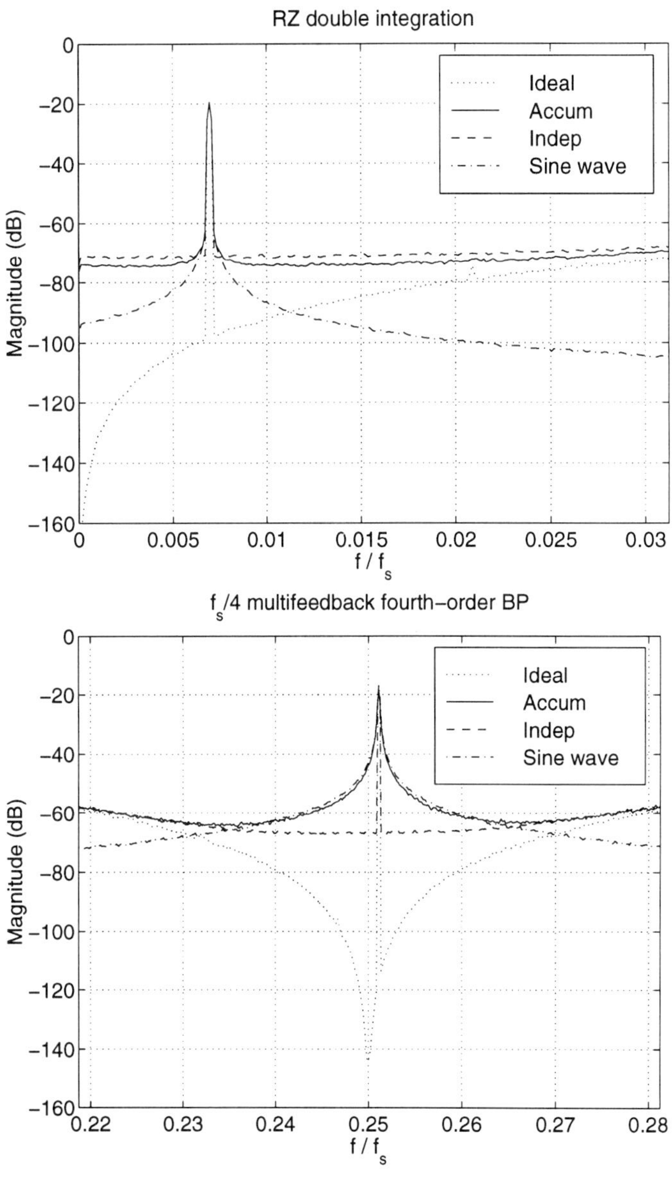

Figure 5.9 (continued). Comparison of $\Delta\Sigma$M spectra.

large tone close to the upper in-band frequency edge will have higher skirts and hence lower peak SNR than if the tone were close to dc. Moreover, the $f_s/4$ BP modulator's performance is affected much more severely than any of the LP modulators, as is apparent in the graphs.

The σ_β used in Figure 5.9 is unrealistically high for a practical VCO; it was used simply as an illustration. In Table 5.2, we have shown how more realistic σ_β values would affect the performance of real high-speed CT $\Delta\Sigma$Ms. We have characterized the DR and peak SNR of four modulators:

- a 1GHz-clocking third-order LP design with NRZ DAC pulses and Butterworth pole placement in the NTF with gain 1.5 and spread zeros;
- a 2GHz fifth-order LP design with NRZ DAC pulses and Chebyshev pole placement in the NTF with gain 1.5 and spread zeros;
- the 3.2GHz double integration modulator published in [Jen95], which has NRZ DAC pulses; and
- the 4GHz fourth-order BP multi-feedback modulator from [Gao98a] with a noise notch at $f_s/4 = 1$GHz.

For each modulator, we used (5.21) to pick a reasonable value of n_c given f_s, and (5.22) to find σ_β. Then, DR and peak SNR were measured from simulation of each modulator at two different OSRs, 32 and 64. The modulators were simulated both with ideal (unjittered) and VCO (jittered) clocks, and the input tone for the LP modulators is close to the upper band edge so that jitter skirts will be most pronounced[1].

Looking at the table, we may make the following comments. The ideal modulators have DR and SNR limited by quantization noise only; for the modulators clocked with a VCO, the question is, does jitter noise impose additional performance limitations? For the double integration modulator, the answer is no: performance is still quantization-noise limited for the realistic value of σ_β used. For high-order modulators and/or high OSRs, the likelihood of being jitter-noise limited increases, as is particularly clear in the fifth-order modulator with OSR $= 64$: more than two full bits of DR are lost at this clock frequency. As well, modulators with center frequencies away from dc suffer more from jitter performance degradation, as we expect from Figure 5.9—note

[1]A quick check of the absolute σ_β values in the table makes them appear suspicious: for example, the 4GHz modulator has $T_s = 250$ps and $\sigma_\beta/T_s = 4.451 \times 10^{-5}$, which means $\sigma_\beta = 11.1$fs. A typical Gbit-rate data generator specs edge jitter at a value of a few ps, two to three orders of magnitude higher than this σ_β. However, it must be remembered that we are using accumulated jitter, (5.20), not white. This means the clock phase over hundreds of cycles wanders significantly relative to a coherent reference; it is trivial to show that after N clock cycles, the phase is a Gaussian random variable with variance $N\sigma_\beta^2$. The σ_β/T_s values in the table are correct for a phase noise given by (5.21).

Table 5.2. Performance effects of real VCO clocking on practical modulators.

Modulator	σ_β/T_s	OSR	Ideal clock		VCO clock	
			DR	Peak SNR	DR	Peak SNR
1GHz 3rd order Butterworth	2.236×10^{-5}	32	73.2dB	73.3dB	69.6dB	69.6dB
		64	93.5dB	93.8dB	90.5dB	89.2dB
2GHz 5th order Chebyshev	3.155×10^{-5}	32	82.6dB	82.8dB	79.8dB	79.1dB
		64	114.5dB	100.7dB	101.6dB	92.1dB
3.2GHz double integration [Jen95]	3.953×10^{-5}	32	62.7dB	56.3dB	62.6dB	56.3dB
		64	78.8dB	71.3dB	79.0dB	71.4dB
4GHz 4th order BP [Gao98a]	4.451×10^{-5}	32	63.2dB	56.4dB	63.2dB	52.7dB
		64	77.5dB	70.2dB	77.3dB	55.3dB

that maximum SNR for the BP modulator is 4dB worse than ideal at OSR $= 32$ and 15dB worse at OSR $= 64$.

An interesting thing happens when we combine the equations in this chapter to derive the maximum-achievable DR for a VCO-clocked modulator with a phase-noise spec given by (5.21). We assume that the in-band noise is completely white; the full calculation appears in §5.A and the result is equation (5.A.10),

$$\text{Maximum DR (bits)} \approx 19 - 0.5 \log_2 f_N, \tag{5.23}$$

where f_N is the Nyquist rate expressed in MHz. This depends *only* on desired conversion bandwidth—it is independent of clock frequency so long as the VCO conforms to (5.21)! As an example, a 32MHz converter has $f_N = 64$MHz, and (5.23) says VCO jitter will limit the performance to no more than 16 bits in this band. But this is far more than the resolution achieved by the fast modulators in Table 3.2. Clearly, VCO phase noise is unlikely to be the limiting factor in a high-speed modulator.

To conclude, the quality of integrated VCOs for cellular radio applications is good enough that the DR of only very high-resolution wideband LP $\Delta\Sigma$Ms would likely be affected. Fast BP designs might be more problematic in terms of the peak SNR lost due to jitter skirts appearing on the output tone. For [Gao98a] which does band pass conversion at 1GHz, we might think we can address this by downconverting to a frequency of a couple hundred MHz and doing the ADC there instead where the jitter skirts in a $\Delta\Sigma$M would be less severe; however, we must remember that the downconversion operation itself must be done with a jittered clock, and this introduces skirts on the desired tone in the mixing process.

4. LATCHES AND METASTABILITY

Published high-speed CT $\Delta\Sigma$Ms tend to be bipolar-only circuits with a one-bit quantizer. A typical quantizer for such circuits was shown in Figure 2.6, reproduced here as Figure 5.10. As we said earlier, the transistors in the dotted box can be reconnected to produce RZ instead of NRZ waveforms. The dashed box contains the four transistors responsible for regeneration: when the circuit is enabled, the voltage difference V_{rd} at the bases of the emitter follower transistors is amplified by positive feedback until the maximum positive (digital $+1$) or negative (digital -1) voltage difference is reached. In an ideal latch there exists a third output state, the *metastable* state, where the inputs are balanced resulting in a 0V differential output signal. This state is unstable in that a slight perturbation (e.g., from circuit noise) will push the latch towards one of its stable states, hence the metastable state itself is never observed in practice.

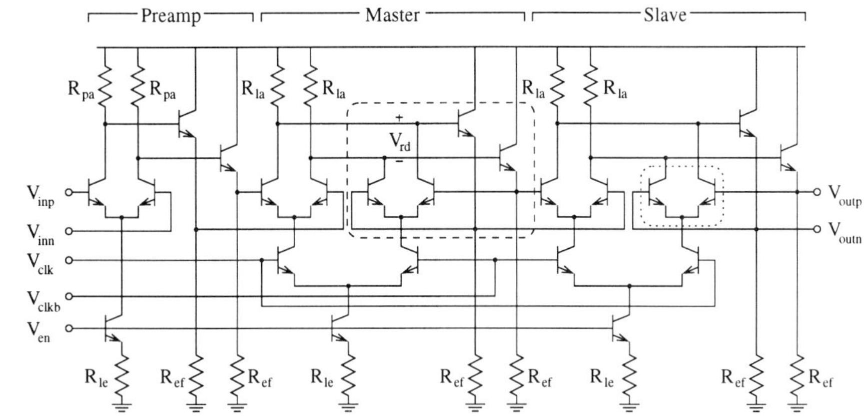

Figure 5.10. Typical high-speed CT ΔΣM master-slave latched comparator with preamplifier.

4.1 DIGITAL CIRCUITS VS. ΔΣMS

The usual analyses of metastability in digital latches [Vee80, Hor89] treat the regenerative circuit as a single-pole system where the voltage difference at $t = 0$ increases exponentially with a time constant inversely proportional to the gain-bandwidth (GB) product of the system. Such a treatment is valid here: Figure 5.11a is a SPICE transistor-level simulation of just the master portion of Figure 5.10 with input voltages given in the legend box. The differential pair amplifies the input voltage in the first half clock cycle, then the regenerative quad is enabled at $T = 1$ns and the value V_{rd} rises exponentially ($\log V_{rd}$ is a straight line) until near the output voltage limit.

In digital circuits, the usual question to be answered is, what is the probability that the latch output is a valid digital level at time t given a certain setup time? In CT ΔΣMs, we are interested in a different question. Figure 5.11b plots the output of a M/S latch whose input is first driven negative to make the latch output -1, then slightly positive to the value in the legend box. Note that the time when the latch output crosses zero on its way to $+1$ varies as a function of the positive input voltage (and that very small positive inputs cause the latch to produce a glitch). This output voltage drives the DACs, and variations in its zero-crossing time (ZCT) have exactly the same effect as quantizer clock jitter—random edge variations modulate out-of-band noise into the signal band and whiten the spectrum. Thus, the question that concerns us is, what is the exact shape of the DAC output waveform? Most particularly, how do its ZCT and rise time vary for input voltages changing sign between clock periods?

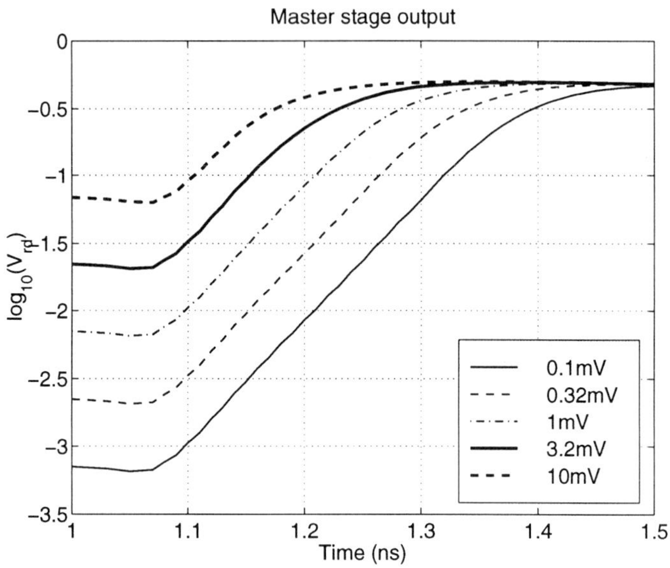

Figure 5.11a. Output of master stage only.

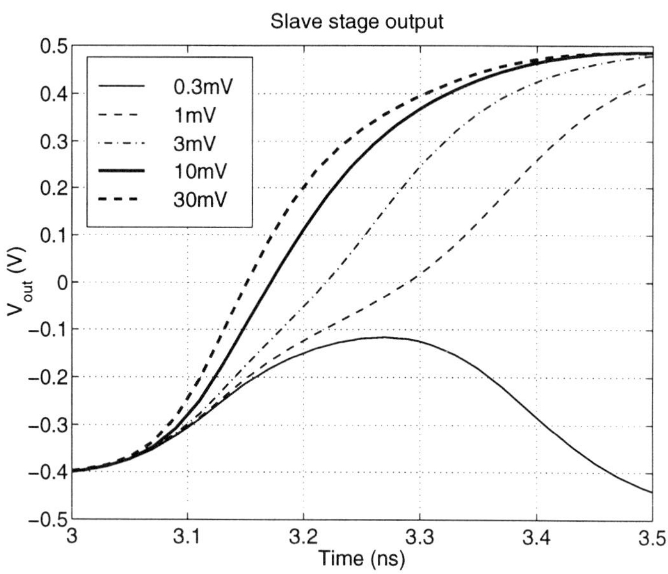

Figure 5.11b. Output of slave stage of a full M/S latch.

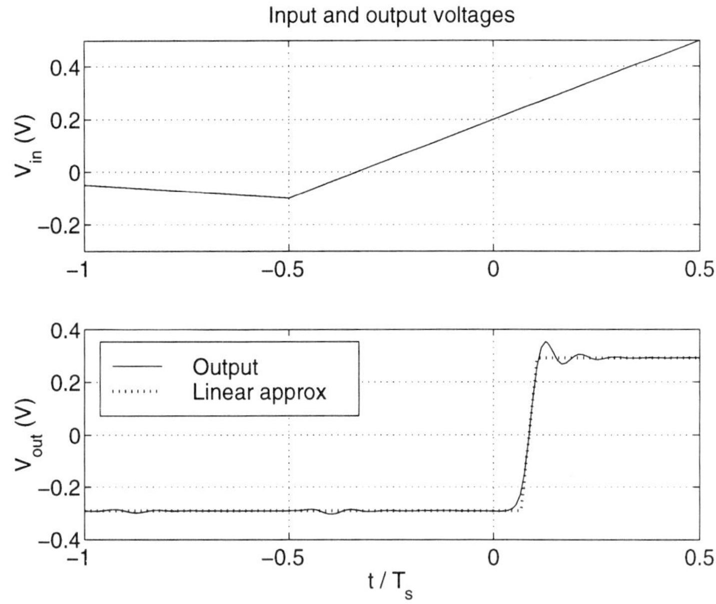

Figure 5.12. SPICE input and output waveforms with linear approximation to output.

4.2 CHARACTERIZATION METHOD FOR ΔΣMS

We determine the ZCT and rise time characteristics of a given latch from simulation. An analytic approach would be preferable, but finding one proved difficult; analyzing the latch output is one thing, but it is followed by a level-shifting follower and a DAC whose behavior must also be characterized. However, we show below that our empirical model works well.

To derive it, a transistor-level SPICE file describing the complete feedback circuit (from latch input to feedback output) is composed. The input to the latch is made to be a piecewise-linear wave which first goes negative to drive the feedback output negative, then positive with slope v_{sl} so that at the next clocking instant the latch input is a specified value v_x. For many different (v_x, v_{sl}) pairs, the ZCT relative to the previous sampling instant (which we call $\tau_d = \rho_d T_s$ for "delay time") and the feedback output rise time (which we call $\tau_r = \rho_r T_s$ for "rise time") are found from simulation. If the circuit were single-ended rather than differential, it would be possible to characterize ρ_d and ρ_r both for rising and falling latch inputs.

Example 5.5: The process is illustrated in Figure 5.12, which is for an M/S latch like Figure 5.10 except with no preamplifier stage. The clock rate is $f_s = 500\text{MHz}$ ($T_s = 2\text{ps}$), and the transistors have a switching speed of about $f_T = 12\text{GHz}$. In the upper graph, we see $(v_x, v_{sl}) = (0.2, 0.6)$

for the input wave V_{in} at $t = 0$. The latch output goes through an emitter follower to a differential pair DAC whose collectors have been terminated with resistors. It is the differential resistor voltage that we plot as V_{out} in the bottom graph. We calculate $(\rho_d, \rho_r) = (0.0860, 0.0386)$ and this is plotted as the dotted line in the bottom graph; the approximation to the actual waveform is quite good.

Our input wave is such that we are characterizing the quantizer by driving it hard one way, which makes it emit a "strong pulse", then weakly the other way, which makes a "weak pulse". Experience shows that this is by far the most common case—rarely is the quantizer input of a real modulator such that the quantizer emit two weak pulses in a row. □

Using Perl [Wal96] helps greatly to automate the procedure for many (v_x, v_{sl}) pairs. Curves for ρ_d and ρ_r for our M/S latch and DACs from Example 5.5 are plotted in Figure 5.13. These curves indicate that for inputs close to zero, both the ZCT and the rise time increase, c.f. Figure 5.11b. Moreover, for small enough inputs, no zero crossing is measured, which is what we saw with the glitch in Figure 5.11b for $V_{in} = 0.3$mV—this is an example of quantizer hysteresis. And, as the input passes through v_x with higher slopes, delayed zero crossings and hysteresis happen for larger values of v_x. These curves have been normalized so that v_x is relative to the expected full-scale quantizer input, which for our example circuit happens to be ± 1V. v_{sl} values are normalized to full scale volts per clock period and swept from 0 to 2. $v_{sl} = 2$ corresponds to a straight line quantizer input that goes from positive to negative full-scale over one clock period. Typically the maximum quantizer slope can be observed to be about half this much, though the quantizer input isn't usually a perfectly straight line (recall x_2 in Figure 3.1). Nonetheless, $v_{sl} = 2$ should be larger than most practically-occurring slopes.

The data from Figure 5.13 is used as input to the RK4 simulation program: at each clocking instant, the program calculates (v_x, v_{sl}) and uses linear interpolation to find (ρ_d, ρ_r), which are then used to set the feedback pulse's delay time and rise time. Essentially, we are employing the technique of *behavioral modeling* [Cur95].

4.3 VALIDATION OF QUANTIZER MODEL

How good is the behavioral model? In other words, how well do the results from our $\Delta\Sigma$M simulator using the behavioral quantizer model agree with those from full-circuit simulation?

Example 5.6: The comparator and feedback circuitry of the prototype double integration modulator in Chapter 2 §4.2 designed in an $f_T = 25$GHz process clocking at $f_s = 1$GHz were characterized as described in Example 5.5, and as many of the parameters from the actual circuit as possible

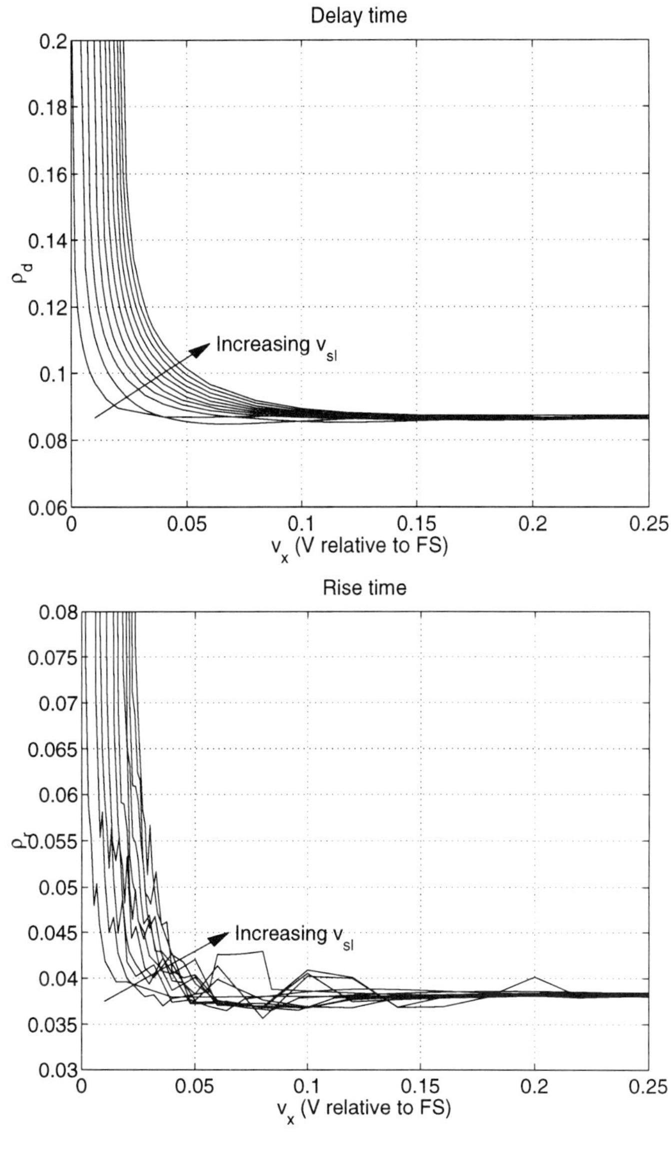

Figure 5.13. Example of numerically-characterized ρ_d and ρ_r values.

(such as finite integrator gain and input resistance) were included in an input file to the RK4 simulator. For a -4dB input at 3.1MHz, output spectra for a 16384-point SPICE simulation and 64 averaged 16384-point periodograms from the RK4 simulation are shown in Figure 5.14a. The spectrum details agree quite well, and there is acceptable agreement between calculated SNR values at OSR $=$ 32 and 64, as shown on the graph. The RK4 program predicts a slightly lower white noise floor due to metastability than SPICE. Figure 5.14b shows a histogram of the quantizer input pdf $p_x(\alpha)$ from each simulator, and good agreement is seen—we are modeling the behavior and voltage levels in the real circuit quite well.

A dynamic range plot is shown in Figure 5.15a. The RK4 values of SNR were found from 32 averaged 4096-point periodograms, and in SPICE each value was found from a single 4096-point simulation. The agreement between the two is quite good, and it is worth noting that while each RK4 simulation of 128k output bits took about 30 seconds, a single 4096-bit SPICE simulation took over four hours. □

Behavioral models are meant to increase simulation speed while maintaining accuracy, and we see that our quantizer model scores well on both counts.

Example 5.7: We made an interesting discovery in the course of this work: the effect of metastability for small input magnitudes. The dashed line in Figure 5.15a has slope 1dB/dB, which is the expected slope of the SNR vs. input magnitude curve. This slope is observed in simulation for large input amplitudes, but as input amplitude decreases, we achieve SNR $=0$ for an input magnitude of -42dB, whereas the dashed line predicts SNR $=0$ at -52dB input. With a -40dB modulator input, observation of the quantizer input as a function of time reveals the behavior shown in Figure 5.15b: up to about sample 160, the quantizer behaves as it should, but then the modulator enters a $\{+1, -1\}$ limit cycle from which it does not escape at a later time. Clearly, the modulator output no longer encodes the input signal at this point. We observed this behavior in both RK4 and SPICE simulations.

It is known that integrators with finite gain can cause such behavior [Fee91]. However, it was found in RK4 simulations with a metastable quantizer that the behavior occurred even with ideal (i.e., infinite-gain) integrators. It was thought that perhaps a quantizer with hysteresis alone (i.e., with *constant* ρ_d and ρ_r) might also cause the $\Delta\Sigma$M to exhibit the behavior, but this was not found to be the case with infinite-gain integrators. Therefore, it appears that quantizer metastability can result in worsened sensitivity of an otherwise ideal CT $\Delta\Sigma$Ms to small input levels.

It should be noted that the metastability is indeed being excited in Figure 5.15b. From samples 160 to 300 or so the bottom envelope of the

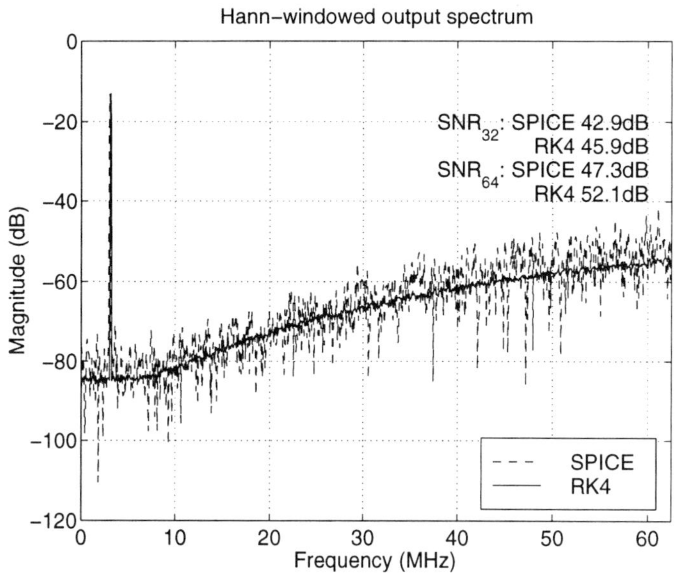

Figure 5.14a. Comparison of SPICE and Runge-Kutta output spectra.

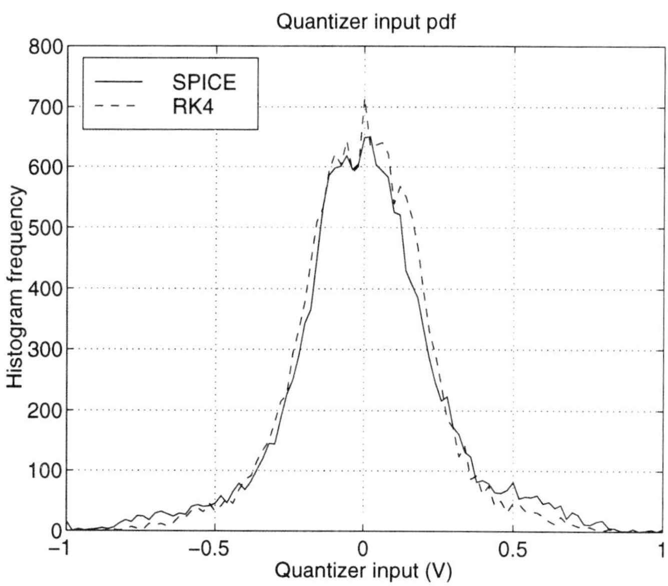

Figure 5.14b. Comparison of SPICE and Runge-Kutta quantizer input pdfs.

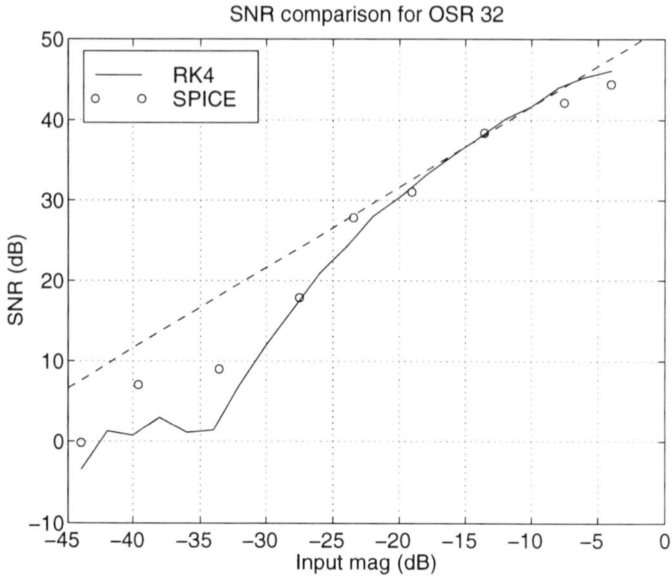

Figure 5.15a. Dynamic range plot in SPICE vs. Runge-Kutta.

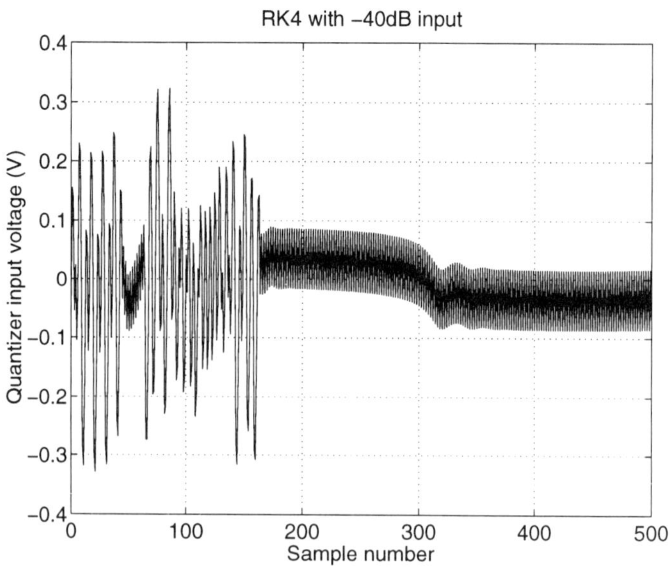

Figure 5.15b. Quantizer input limit cycle for low input magnitude.

quantizer input is near zero, which activates the metastability. It "escapes" from this mode of behavior only to have the *top* of the envelope approach zero and activate the metastability at sample 320. The metastability is excited alternately by the envelope top and bottom every few hundred clock cycles. □

This result is of grave significance because it implies using a single large-amplitude tone to estimate modulator resolution is insufficient: one might predict a DR based on an incorrect assumption of a slope of 1dB/dB down to SNR = 0. To the authors' knowledge, this is a previously unpublished result.

Admittedly, there is no absolute guarantee that changing the quantizer circuit will mean the agreement between behavioral and SPICE simulation remains good. However, we have more than one reason to be confident that our behavioral model has identified the key issues. First, we have good agreement not just on SNR, but on output spectrum details and quantizer input pdf too. Second, the behavioral model correctly predicted the limit cycle behavior, a result which was *a priori* unknown. For the remainder of this chapter we use only the behavioral model for circuit performance measurements.

5. REAL QUANTIZER PERFORMANCE EFFECTS

The design of CT $\Delta\Sigma$Ms is usually done assuming an ideal quantizer, which has no hysteresis and makes a decision instantly. The characteristics of such a quantizer are plotted in Figure 5.16a: ρ_d is always zero no matter how small v_x is. Practical quantizers suffer from three nonidealities which can be distinguished as follows.

1. Excess delay, Figure 5.16b, means a vertical shift of the quantizer curve; more excess delay means a greater vertical shift. As we noted in Chapter 4, the effect of excess delay is to increase in-band noise and lower MSA which in combination compromise overall modulator DR.

2. Hysteresis, Figure 5.16c, causes a horizontal shift of the curve proportional to the amount of hysteresis. From §1.3, quantizer hysteresis in an otherwise ideal system adds noise power in the baseband for LP modulators [Bos88]. An example of its effects for the third-order LP Butterworth $\Delta\Sigma$M studied earlier is depicted graphically here. In Figure 5.17a, we see $p_x(\alpha)$ becoming wider with hysteresis; this is expected because as long as the quantizer output bit remains the same, the circuitry inside the loop will continue integrating in the same direction, enlarging signal swings. In a modulator whose integrator outputs clip, hysteresis introduces harmonic distortion; moreover, too large an internal signal excursion range leads only to *gradual* instability and hence DR loss, as shown in Figure 5.17b—hardly any performance is lost even for large hysteresis. By contrast, [Cha92] found 1%

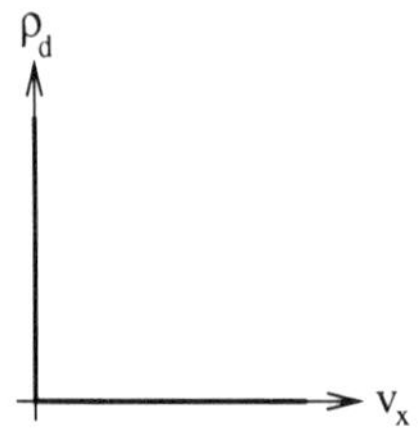

Figure 5.16a. Ideal quantizer characteristic.

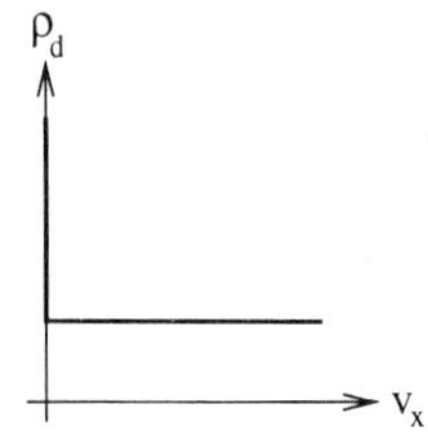

Figure 5.16b. Quantizer with excess delay.

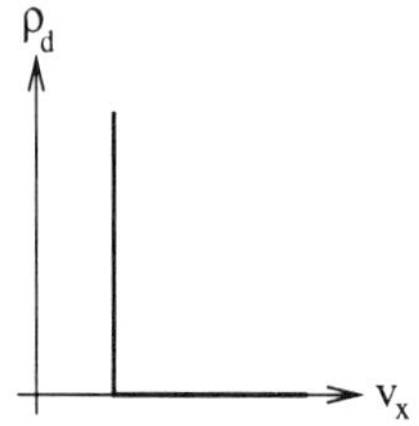

Figure 5.16c. Quantizer with hysteresis.

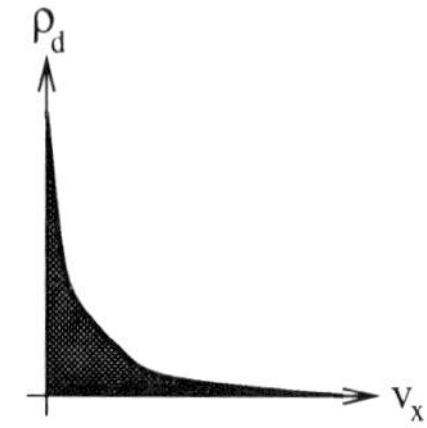

Figure 5.16d. Quantizer with metastability.

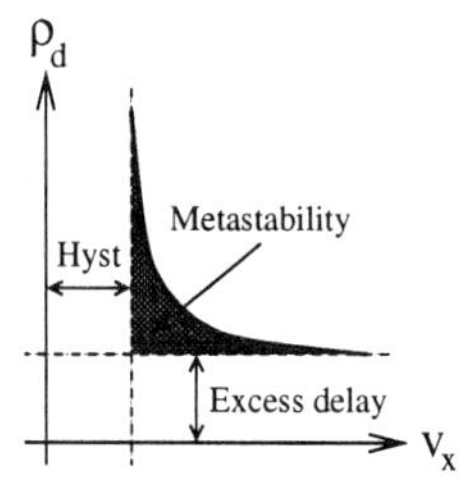

Figure 5.16e. Practical quantizer.

hysteresis caused significant performance loss, though this is at odds with other publications and our own results.

3. Metastability, Figure 5.16d, means that the sharp corner in the ideal quantizer characteristics becomes smooth instead. The severity of the metastability is related to the amount of area underneath the curve: curves with a more abrupt corner have less area under them and hence less metastability.

A real quantizer has all three effects simultaneously, Figure 5.16e, as we saw in Figure 5.13. Along with the losses already caused by excess delay and hysteresis, metastability introduces two additional performance-limiting effects. First, at low input amplitudes, there is the output limit cycle behavior mentioned in Example 5.7 and depicted in Figure 5.18a for a double integration modulator. Second, at higher input amplitudes spectral whitening occurs due

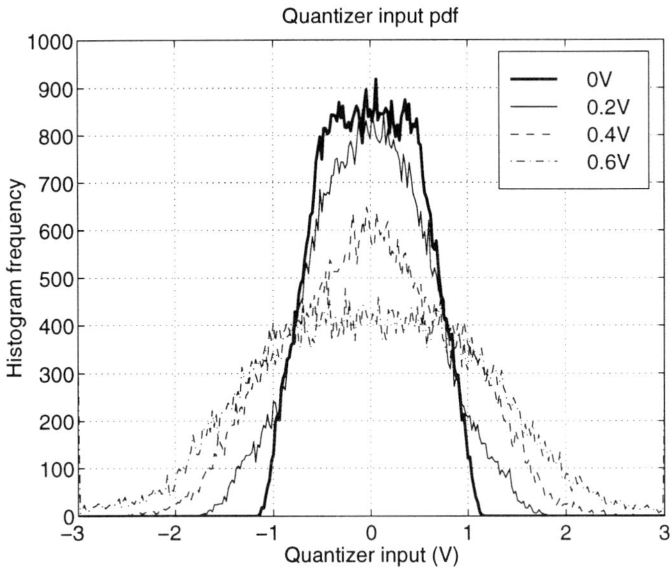

Figure 5.17a. Effect of quantizer hysteresis on quantizer input pdf in third-order Butterworth modulator.

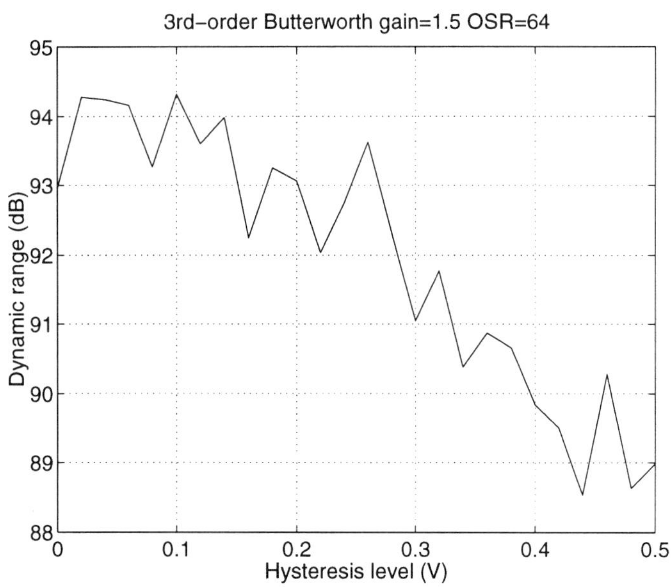

Figure 5.17b. Effect of quantizer hysteresis on DR in third-order Butterworth modulator.

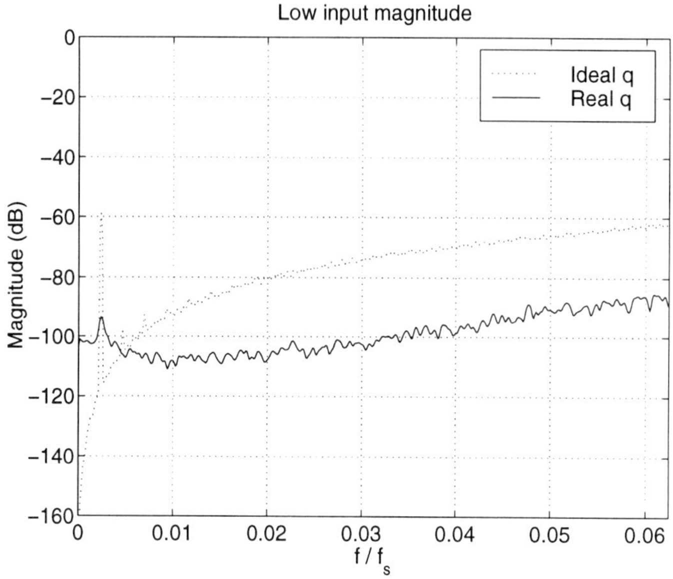

Figure 5.18a. Typical output spectrum in double integration ΔΣM with a real quantizer and a low-amplitude input signal.

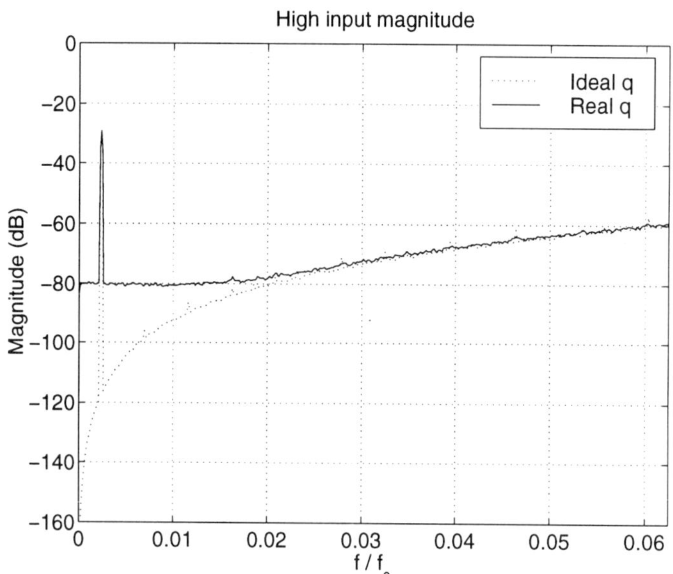

Figure 5.18b. Typical output spectrum in double integration ΔΣM with a real quantizer and a high-amplitude input signal.

to the *variability* of ρ_d mentioned in Chapter 2 §4.2. A typical spectrum is depicted in Figure 5.18b.

Example 5.8: The DR impact of using a quantizer with the characteristics in Figure 5.13 on several kinds of LP $\Delta\Sigma$M is shown in Figure 5.19. In order to make the comparison fair, the modulators had their feedbacks scaled so that they all had the same quantizer input pdf standard deviation of $\sigma_x = 1/3^2$. We observe the following:

- An ideal modulator exhibits $6m + 3$dB/oct improvement of SNR with OSR, where m is the modulator order. A modulator with a metastable quantizer will, for large enough OSR, be limited to a mere 3dB/oct improvement because the noise notch in the output spectrum is filled in with white noise. For the quantizer from Example 5.5, going from OSR $= 32$ to OSR $= 64$ (shown by the dash-dot lines) gives us only 3dB SNR improvement. This means the DR at OSR $= 64$ for all modulators is cut drastically by a full factor of two.
- The dashed lines show modulators with hysteresis and fixed

$$(\rho_d, \rho_r) = (8.6\%, 3.9\%);$$

compared with the ideal modulator, DR is hardly compromised at all, perhaps 0–3dB depending on the modulator order.
- The limit cycle behavior in $\Delta\Sigma$Ms with metastable quantizers mentioned earlier seems only to affect the lower-order modulators: the higher-order modulators have 1dB/dB slope all the way down to low input magnitudes, except perhaps for a slight dip near -35dB.

Metastability clearly has a major impact on the DR of these high-speed CT $\Delta\Sigma$Ms. □

There is a relationship between performance lost by metastability and that lost by clock jitter which can be explained as follows.

Example 5.9: We know that random variation in DAC pulse width (DPW) fills the output spectrum noise notch with white noise. The solid line in Figure 5.20a shows a histogram of DPW variation for the fifth-order modulator with an ideal sampling clock and a metastable quantizer with characteristic given by Figure 5.13. The standard deviation of this distribution is $\sigma_{ms} = 5.95 \times 10^{-3} T_s$. To get the same DPW variance from

---

[2] Since $p_x(\alpha)$ is roughly Gaussian [Bos88], fixing σ_x assures roughly the same distribution of abscissae in the quantizer characteristic in Figure 5.13, and hence a roughly-comparable ordinate distribution.

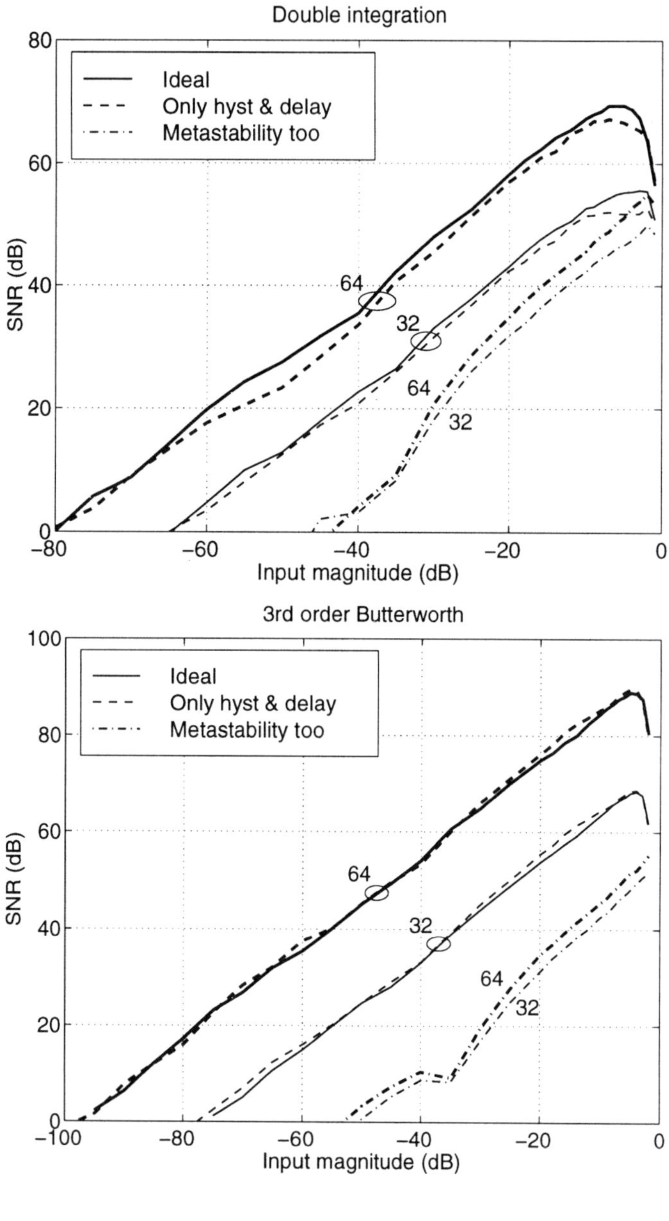

Figure 5.19. Dynamic range plots. Numbers on curves are OSR values.

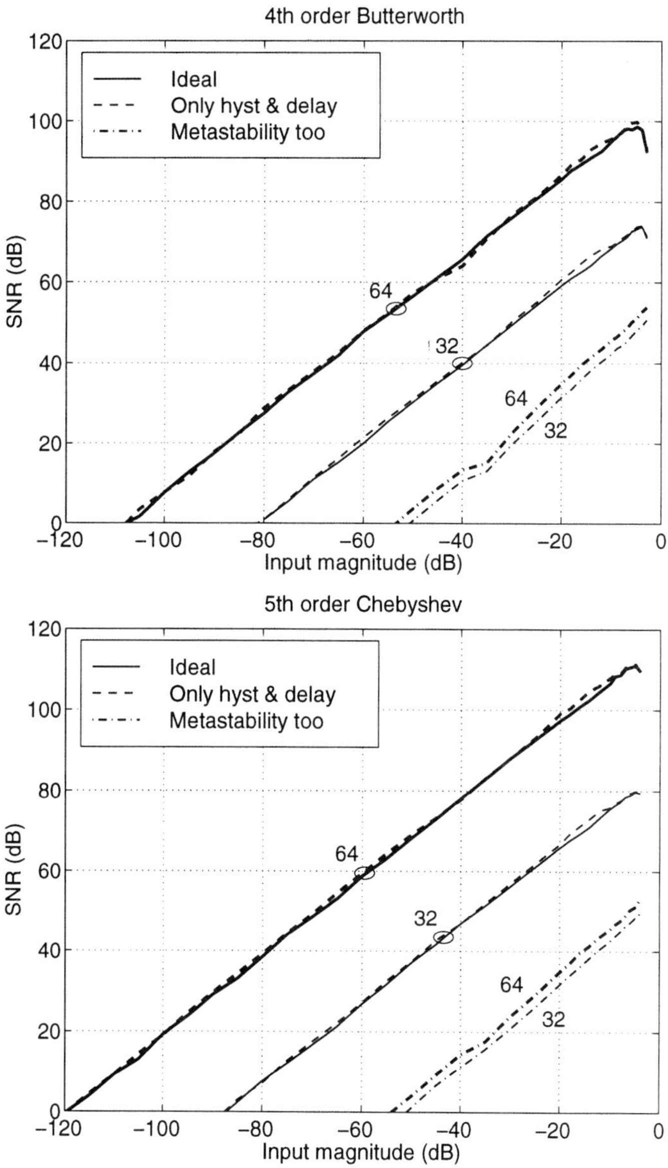

Figure 5.19 (continued). Dynamic range plots.

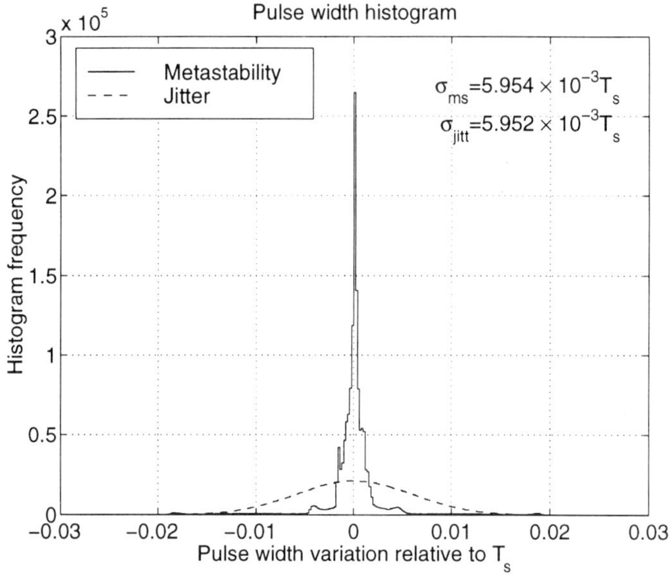

Figure 5.20a. DAC pulse width variation histogram for clock jitter vs. quantizer metastability.

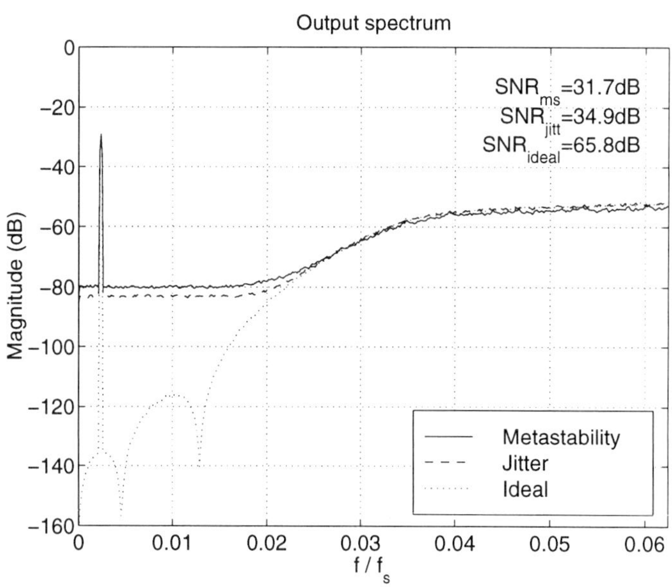

Figure 5.20b. Output spectrum for clock jitter vs. quantizer metastability.

a modulator with an ideal quantizer and a clock with independent jitter, we must set the jitter variance to

$$\sigma_\beta = \sigma_{ms}/\sqrt{2}. \tag{5.24}$$

This results in the dashed-line histogram in Figure 5.20a. Note that $\sigma_{jitt} \approx \sigma_{ms}$. Since the DPW variance is about the same in both cases, §2. taught us that the spectrum whitening should also be about the same. Figure 5.20b illustrates this to be the case: the SNR value for the modulator with the metastable quantizer is 31.7dB, while that for the modulator with clock jitter is 34.9dB. The noise floors are close but not identical because the distribution of the DPW histogram for a metastable quantizer is not particularly Gaussian. Even still, the agreement is quite good. □

Clock jitter is not identical to metastability because clock jitter does not cause the limit cycle behavior observed in Example 5.7 for low input amplitudes; it is the white noise levels that are roughly the same in both for large-enough input amplitudes. Though the DRs might differ, the peak SNR measured in both would come out about the same.

6. MITIGATING METASTABILITY PERFORMANCE LOSS

How can we overcome the performance penalties imposed by quantizer metastability? Several answers to this question suggest themselves when we consider the source of the loss: the variations in the DPW caused by finite quantizer regeneration speed. We observed at the start of §5. that this variance σ_{DPW} is related to the area under the metastability curve in Figure 5.16e, or equivalently the "sharpness" of the corner in the curve. What approaches might we take to reduce its area or increase its sharpness?

6.1 PARAMETER SCALING

The first thing we might think of is to scale the modulator parameters to enlarge the quantizer input standard deviation σ_x. This works as follows: we know that if the magnitude of the quantizer input is small, then the ZCT increases. For the quantizer in Figure 5.13, inputs which cause increased ZCT are approximately those for which $|v_x| < 100\text{mV}$. By increasing σ_x, we widen the range of possible quantizer inputs so that the probability of $|v_x| < 100\text{mV}$ is decreased. Alternately, we may think of this as scaling the v_x axis by compressing the metastability curves towards the ρ_d axis. This reduces the effective area under a given curve and hence reduces σ_{DPW}.

Example 5.10: Example 5.8 used $\sigma_x = 0.33$. We illustrate the effect of choosing σ_x ranging from 0.1 up to 0.5. In Figure 5.21a, the metastability

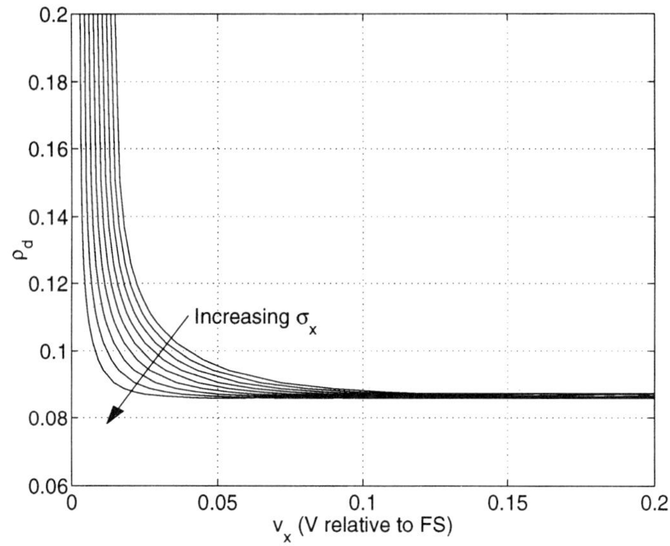

Figure 5.21a. Quantizer metastability curves as a function of σ_x.

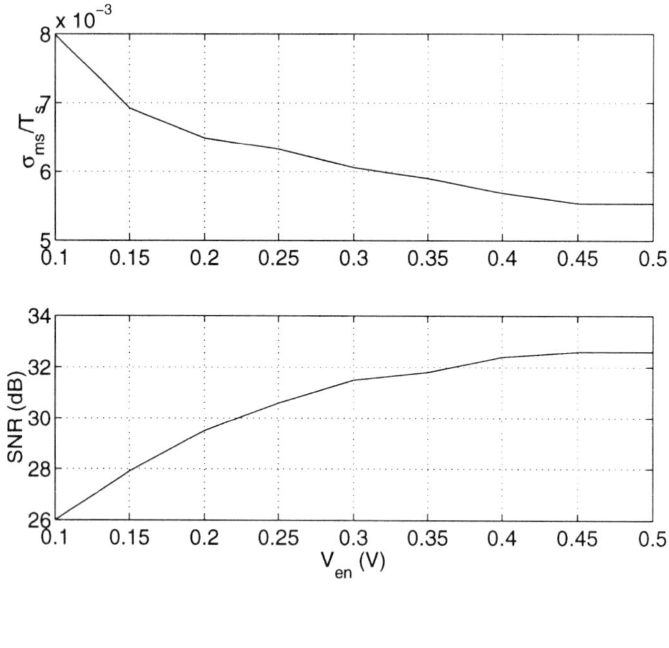

Figure 5.21b. Modulator performance as a function of σ_x.

curve for $v_{sl} = 0.6$ is plotted as a function of σ_x; we can see the corner of the curve becoming sharper as σ_x increases, which leads to more favorable σ_{ms}/T_s and SNR in Figure 5.21b. □

Generally, it is a good idea to have the quantizer input span as large a range as possible. The range can be increased by, for example, choosing smaller integrating capacitors and larger feedback currents. Circuit constraints will ultimately limit the maximum achievable range; for our case, the quantizer input can swing about ± 1V differential while still keeping all transistors operating in their forward active regions. Having too small a swing range is to be avoided because as we see in Figure 5.21b it quickly becomes detrimental for SNR.

6.2 REGENERATION TIME

In §4.1, we said that the regeneration time of a latch is inversely proportional to the GB product of the regeneration circuit. If we were to increase this GB product, the corners of the metastability curves would become sharper as follows: the slope of the curves in Figure 5.11a would increase, which in turn would, we hope, mean that it would take a *smaller* input level v_x for the curves in Figure 5.11b to exhibit increased delay—in other words, for the set of v_x values in that legend box, the rising output edges would be bunched more closely together. In turn, the curves in Figure 5.13 would rise towards infinity more abruptly, i.e., the corner becomes more pronounced.

Example 5.11: One way to affect the regeneration time constant τ_{rg} of the latch in Figure 5.10 is to change the current in the regenerative quad; this is accomplished by altering the voltage V_{en}. To keep the comparison reasonable, we will adjust V_{en} and R_{la} simultaneously to keep the latch output voltage swing at around ± 300mV. Sweeping V_{en} over the range 0.90V to 1.35V and extracting τ_{rg} from simulation yields the curve of Figure 5.22a: as is usual with ECL circuits, there exists a current/load resistor combination which minimizes rise time which for this latch occurs at about $V_{en} = 1.15$V. While fast linear settling is important, surely as important for ZCT is the nonlinear settling behavior, i.e., slewing. Figure 5.22b plots the slew rates at both the latch output (top) and DAC output (bottom) as a function of V_{en}.

Figure 5.23a shows quantizer metastability curves for $v_{sl} = 0.6$ as V_{en} changes. There is some sharpening of the corner with increased V_{en}, but as an added bonus also lower ρ_d and hysteresis. Just as diminishing returns are apparent there, so too are they in Figure 5.23b where the SNR is plotted against V_{en}. The optimum SNR does not quite occur where τ_{rg} is a minimum—very minor improvements are obtained as V_{en} is raised further because of the increase in slew rate. □

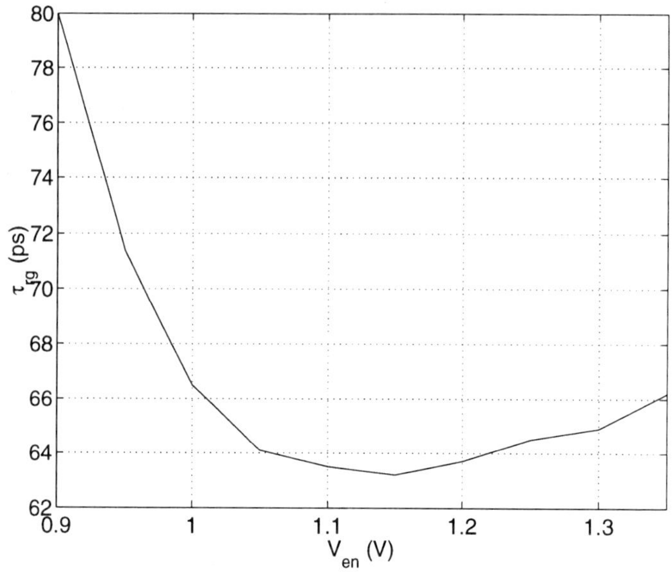

Figure 5.22a. Effect of V_{en} on regeneration time.

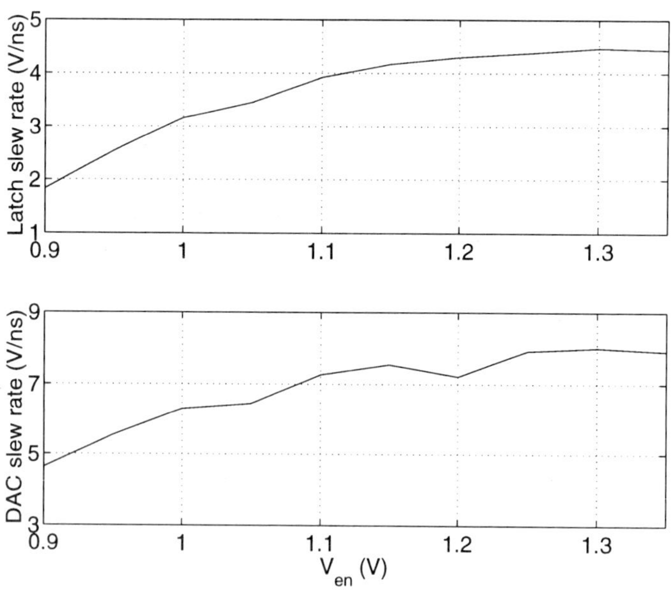

Figure 5.22b. Effect of V_{en} on slew rate at latch and DAC outputs.

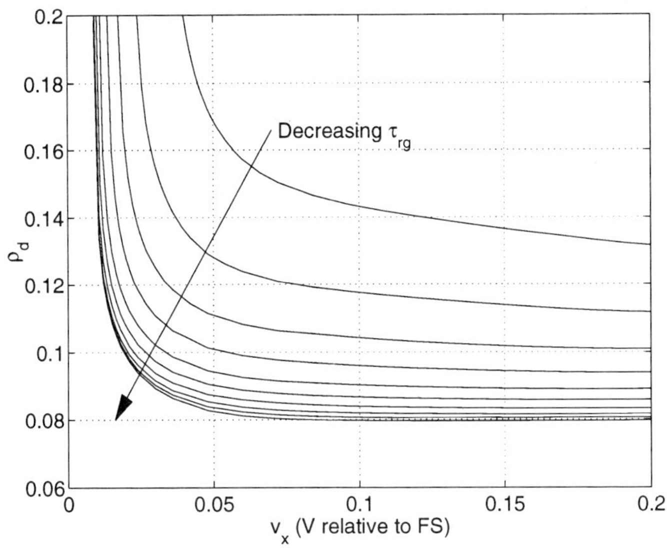

Figure 5.23a. Quantizer metastability curves as a function of τ_{rg}.

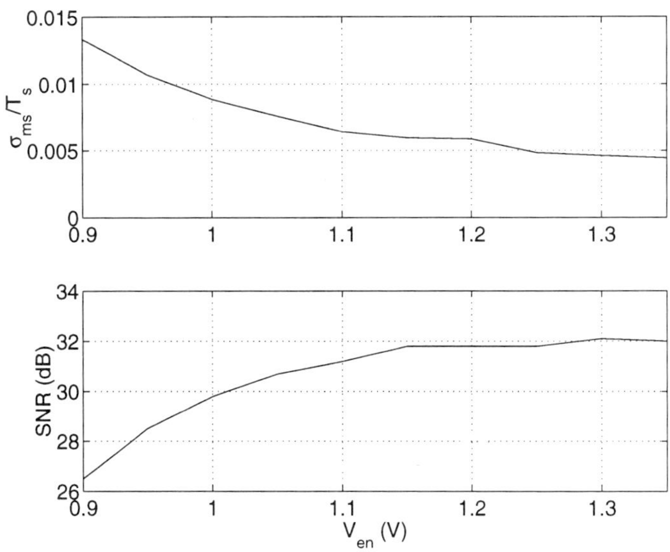

Figure 5.23b. Modulator performance as a function of τ_{rg}.

Ensuring adequate regeneration is a good idea in a $\Delta\Sigma$M's latch. Setting V_{en} very high might use more power than necessary to achieve a given SNR, so there exists a tradeoff between power consumption and SNR, though we could use smaller transistors whose peak speed occurs at lower bias current.

6.3 PREAMPLIFICATION

A third thing we can try is using a latch with a preamplifier. This is similar to signal scaling in §6.1 but not identical because we actually insert a new circuit element into the forward path. How do we choose its gain? Traditional analyses show that for cascaded amplifiers there exists an optimum gain per stage that maximizes the overall amplifier GB product and hence the amplifier speed. Depending on the assumptions made, the optimum gain is either $e = 2.72$ [Sne96, Chap. 2] or $\sqrt{e} = 1.65$ [Lee98, Chap. 8]. In the present circuit, we will consider only one preamplifier stage and study how its gain affects metastability performance in the following example.

> **Example 5.12**: Figure 5.24a shows metastability curves for a quantizer with a simple differential pair preamplifier as depicted in Figure 5.10. The gain was varied between 1 and 10 by changing the value of R_{pa}. Increasing the gain has the desirable effect of sharpening the corner of the metastability curves for constant quantizer inputs (see the solid lines in the figure), but for *slewing* inputs little sharpening can be seen as gain rises. Hysteresis increases slightly with gain, and in fact it has increased substantially over Figure 5.23a from about 10mV to about 30mV. This is not terribly detrimental to performance as we learned in Figure 5.17b.
>
> Figure 5.24b shows that a preamplifier does offer some SNR improvement (about 2dB) over parameter scaling, Figure 5.21b, and regeneration time lowering, Figure 5.23b. There is little point in using a gain above 4, it appears. □

Thus, preamplifying is somewhat beneficial for performance. An ancillary benefit of a preamplifier with an emitter follower buffer between it and the latch input is a reduction in clock feedthrough noise [Lee92].

6.4 ADDITIONAL LATCHING STAGES

A fourth thing we can try is using additional latching stages after the slave stage in Figure 5.10 [Jen95]. Clocking each stage on the opposite clock phase from the previous stage gives the previous stage a good deal of time to settle. The drawback is, each latching stage adds an additional half-sample delay in the feedback loop, and from Chapter 4 this delay is detrimental to stability and dynamic range. However, we can somewhat overcome these problems by tuning the k_r and k_h feedback parameters.

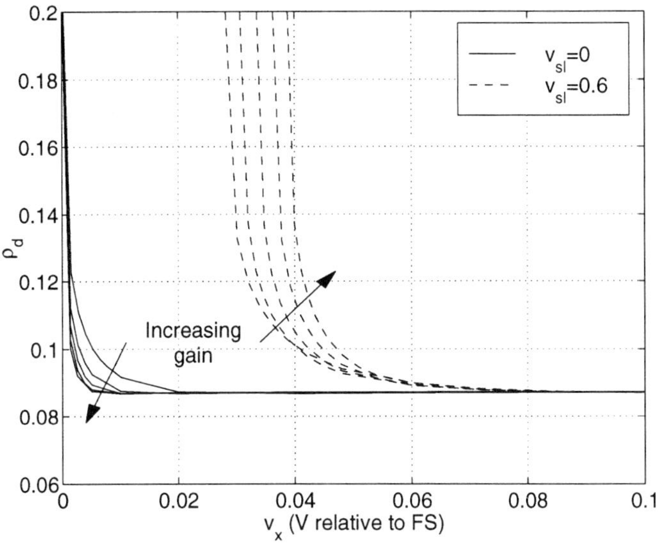

Figure 5.24a. Quantizer metastability curves as a function of preamplifier gain.

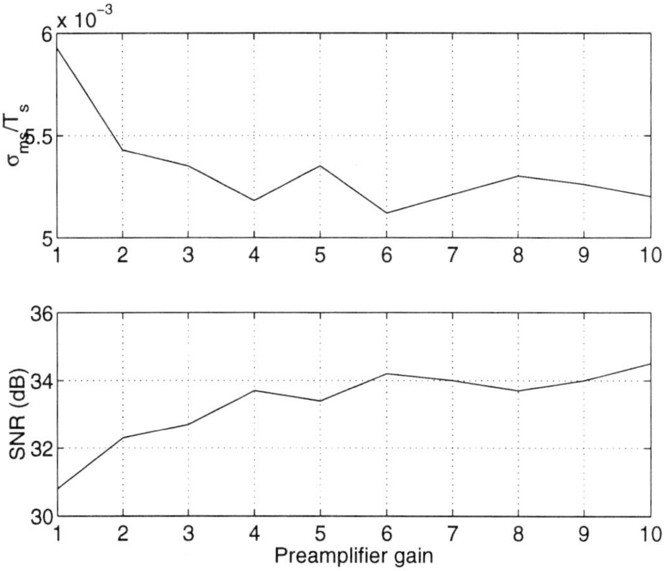

Figure 5.24b. Modulator performance as a function of preamplifier gain.

Example 5.13: Figure 5.25a shows quantizer metastability curves for our baseline latch which has only master and slave half-stages, and a latch that has a third half-stage following the slave which is clocked on the same phase as the master. We have added one half sample of extra delay as can be seen on the right y axis where ρ_{d3} is 0.5 more than ρ_{d2} on the left axis, but the variation of ρ_{d3} with v_x is drastically reduced. This results in a huge improvement—about 40dB—in the white noise floor of the output spectrum, Figure 5.25b. From simulation, we find DPW variance has dropped nearly two orders of magnitude, from $5.9 \times 10^{-3} T_s$ to $1.4 \times 10^{-4} T_s$.

The fifth-order modulator was unstable with $\rho_d = 0.6$ and nominal k values, but the ks were tuned so that the modulator was stable and the DR of a modulator with an ideal quantizer was maximized at $\rho_d = 0.6$. One artifact of the large ρ_d is the peak in the spectrum at $0.025 f_s$, something which is caused by the movement of the equivalent DT NTF poles toward the unit circle as excess loop delay increases. □

We seem to have come across a solution to the metastability problem. How well does it work in general?

Example 5.14: Figure 5.26 shows DR plots for several LP ΔΣMs for an ideal quantizer and the two quantizers with the metastability curves in Figure 5.25a. With two half-latches, there is only about 10% excess delay, but the third half-latch pushes that up to 60%. A modulator with an ideal quantizer and 60% delay usually requires k tuning to remain stable, and even then, the DR is less than for the 10% delay case. However, when the ideal quantizer is replaced with a metastable one, the three half-latch quantizer is the clear DR winner.

Table 5.3 summarizes the results for four cases: an ideal quantizer, a quantizer with two half-latches, an ideal quantizer with 60% excess loop delay and tuned ks to maximize DR, and a quantizer with three half-latches and the same tuned ks. For high-order modulators with high OSR, the white noise resulting from metastability is what limits the achievable performance, though much less severely with three rather than two half-latches. □

The idea of using a third latching stage "to provide additional regeneration" is mentioned in passing in [Jen95, p. 1123], though the claim that "the extra 1/2 delay does not have any impact on modulator performance" seems suspect given the first two rows of Table 5.3: we find a half-sample delay costs 12dB of DR in an ideal second-order CT ΔΣM. Adding a fourth latching stage will not usually be possible because stabilizing a CT ΔΣM with a full sample of delay probably cannot be accomplished through feedback tuning, but even the third stage is clearly highly advantageous for performance.

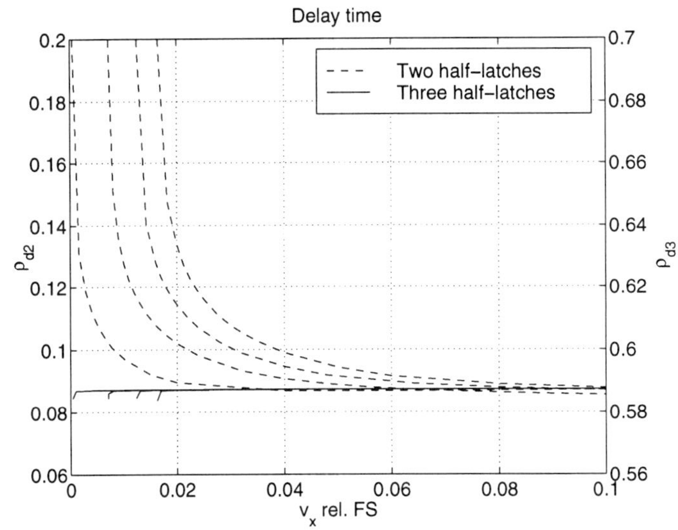

Figure 5.25a. Quantizer metastability curves for a quantizer with an additional latching stage.

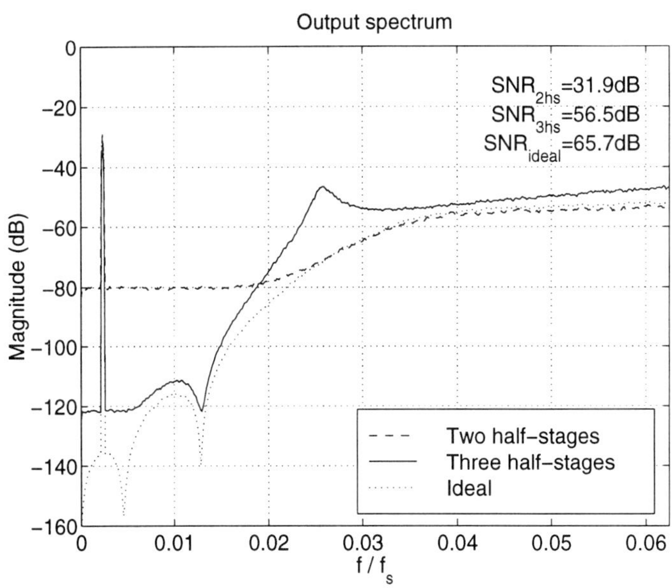

Figure 5.25b. Output spectrum for a quantizer with an additional latching stage.

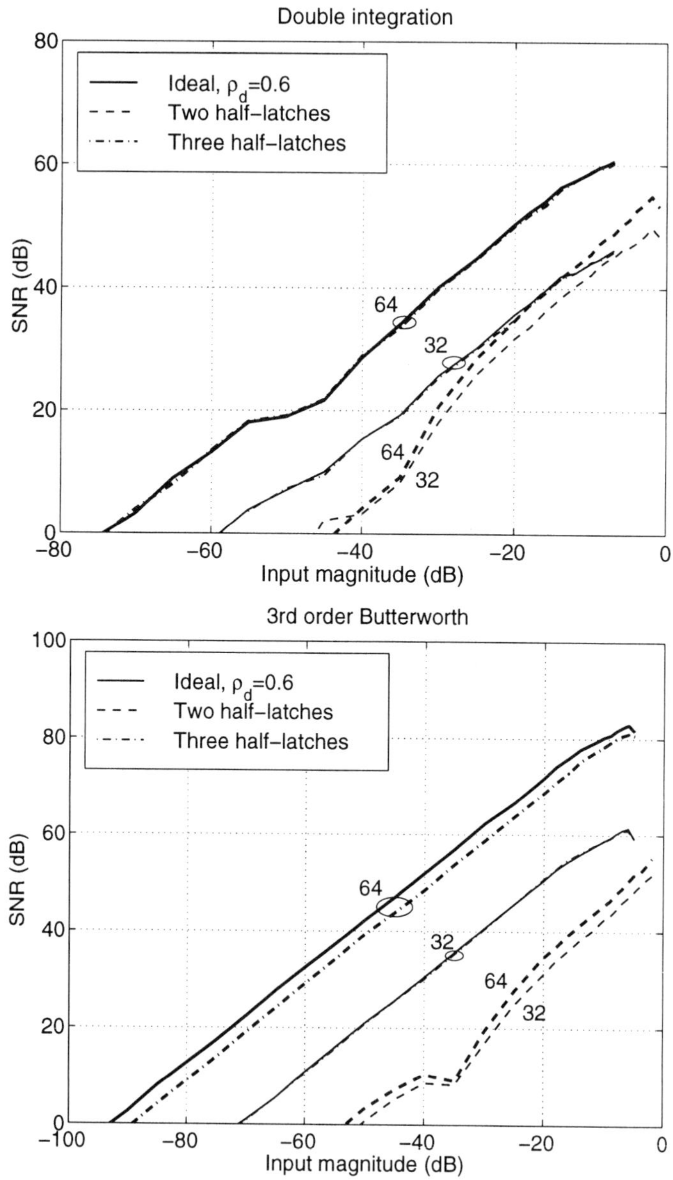

Figure 5.26. Dynamic range plots. Numbers on curves are OSR values.

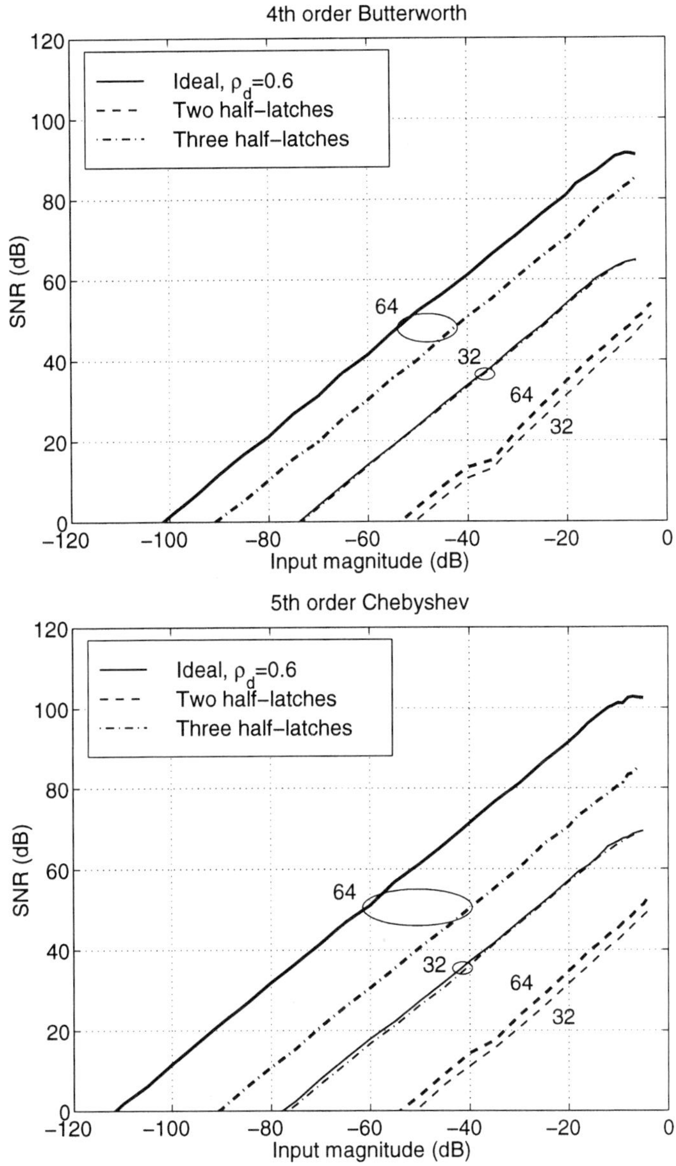

Figure 5.26 (continued). Dynamic range plots.

Table 5.3. Performance of LP modulators with two- and three-half latches against ideal.

Modulator	OSR	Ideal q., $\rho_d = 0.0$		Two half-latch		Ideal q., $\rho_d = 0.6$		Three half-latch	
		DR	Peak SNR	DR	Peak SNR	DR	Peak SNR	DR	Peak SNR
Double integration	32	64dB	56dB	45dB	50dB	52dB	46dB	52dB	46dB
	64	79dB	69dB	43dB	55dB	67dB	61dB	67dB	60dB
3rd order Butter	32	74dB	68dB	49dB	52dB	66dB	61dB	66dB	61dB
	64	95dB	89dB	51dB	55dB	88dB	83dB	84dB	81dB
4th order Butter	32	78dB	74dB	48dB	51dB	68dB	65dB	67dB	65dB
	64	105dB	99dB	50dB	54dB	95dB	91dB	85dB	85dB
5th order Cheby	32	83dB	80dB	47dB	50dB	73dB	69dB	72dB	69dB
	64	116dB	111dB	50dB	53dB	107dB	103dB	85dB	85dB

6.5 OTHER MODULATOR ARCHITECTURES

The previous four subsections assumed LP $\Delta\Sigma$Ms with NRZ DACs; what will happen in LP modulators with RZ DACs and BP modulators which use both RZ and HRZ DACs? An RZ DAC will be affected by metastability in much the same way as an NRZ DAC: the time when the rising edge begins will vary depending on v_x and v_{sl}. To make the latch output return to zero at $0.5T_s$ simply requires connecting the bases of the transistors in the dotted box in Figure 5.10 to the same node as their respective collectors; thus, the falling RZ edge is not affected by metastability. But given the same σ_{ms} in an NRZ vs. an RZ system, it will cause 3dB more noise in the RZ output spectrum, because the same σ_{ms} appears twice as large relative to an RZ pulse (which is half the width of an NRZ pulse).

That being said, applying the ideas of the previous subsection is still worthwhile. In (5.15), we found the equivalent CT loop filter for a double integration modulator when the DAC has RZ pulses. We can do the same thing for a DAC with HRZ pulses, which results in

$$H(z) = \frac{-2z+1}{(z-1)^2} \leftrightarrow \begin{cases} \hat{H}(s) = \frac{-2.5s-2}{s^2}, & \text{RZ DAC} \\ \hat{H}(s) = \frac{-3.5s-2}{s^2}, & \text{HRZ DAC.} \end{cases} \tag{5.25}$$

We can build an HRZ DAC with an additional latching stage in Figure 5.10—and we learned in §6.4 that adding such a stage greatly reduces DPW jitter caused by metastability. Thus, rather than building LP $\Delta\Sigma$Ms with RZ DACs, it behooves us to choose HRZ DACs.

BP modulators with a noise notch at $f_s/4$ have the property that there is a two-sample delay in the numerator of the loop filter $H(z)$, we learned in (4.25). If we choose the one digital delay BP architecture, then we need to insert a full sample of delay in the feedback path. This can be accomplished by using not one but *two* additional half-latches in Figure 5.10—and once again, the previous subsection showed adding half-latches provides immunity to metastability DPW jitter. Thus, any modulator with a two-sample delay in the numerator (which is the case for the BP $f_s/4$ modulators treated here) should be built with one digital delay in the feedback path. In *any* modulator with only a single sample of delay, HRZ-style DAC pulses are called for, as we noted in the previous paragraph for LP modulators.

Using a multibit quantizer is intriguing because it appears to give a win: for an M-level quantizer, there are now $(M-1)$ regions around which metastability can occur but the distance between steps is smaller by a factor of $(M-1)$. DPW variance power is related to the square of this latter quantity, so it offsets the increase in number of metastability regions and appears to result in $10\log_{10}(M-1)$ dB smaller white noise power. This topic could benefit from future study.

7. MAXIMUM CLOCKING FREQUENCY

In our LP NRZ examples so far, we have been clocking at $f_s = 500$MHz in a $f_T = 12$GHz technology. A natural question arises: what is the maximum f_s at which it is safe to clock given a converter resolution specification?

As we increase f_s, two things together limit resolution. First, the transistor switching time starts to become a larger fraction of a clock period. This means the excess loop delay ρ_d and DAC pulse rise time ρ_r start to increase. There comes a point when excess loop delay makes the modulator completely unstable and impervious to stabilization through feedback coefficient tuning. Second, the metastability behavior of even the three half-latch comparator will start to degrade[3]. We study both of these issues in the following example.

Example 5.15: Figure 5.27a shows how ρ_d and ρ_r change as we increase the clock speed of the three half-latch comparator from 500MHz to 2.5GHz. Returning to our crude formula for ρ_d in (4.2),

$$\rho_d \approx \frac{n_t f_s}{f_T}, \tag{5.26}$$

we see that it is somewhat pessimistic. First of all, ρ_d and ρ_r are both quite linear with f_s/f_T, as predicted by (5.26). Moreover, our circuit has $n_t = 3$ transistors in the feedback path (two followers and a differential pair), and so from (5.26) we expect them to have switched fully after

$$0.5 + \frac{n_t f_s}{f_T} = 0.5 + 3\frac{f_s}{f_T} \tag{5.27}$$

where the 0.5 is for the extra half-latch. In terms of ρ_d and ρ_r, full switching is accomplished after

$$\rho_d + 0.5\rho_r, \tag{5.28}$$

c.f. Figure 5.12. Figure 5.27b plots both (5.27) and (5.28); rather than using a factor of $n_t = 3$ in (5.27), a better fit is obtained with a factor of 2.25.

In any case, Figure 5.28a shows σ_{ms} as a function of f_s with both variables on a log scale. This was found from simulating the fifth-order LP NRZ modulator with feedbacks tuned for optimal DR at the given clock frequency over the range 500MHz to 2.5GHz, and finding the variance of the DPW histogram like the one in Figure 5.20a. Using those same simulations with

[3] In §6.5 we suggested using half rather than full DAC pulses for LP modulators. While RZ LP $\Delta\Sigma$Ms remain stable for more excess delay than NRZ LP $\Delta\Sigma$Ms, the worsening of performance due to metastability increases proportionally with f_s for both styles of modulator. As the following example shows, it is metastability that limits DR more than stability; thus, RZ $\Delta\Sigma$Ms have no performance advantage over NRZ $\Delta\Sigma$Ms at high clock speeds.

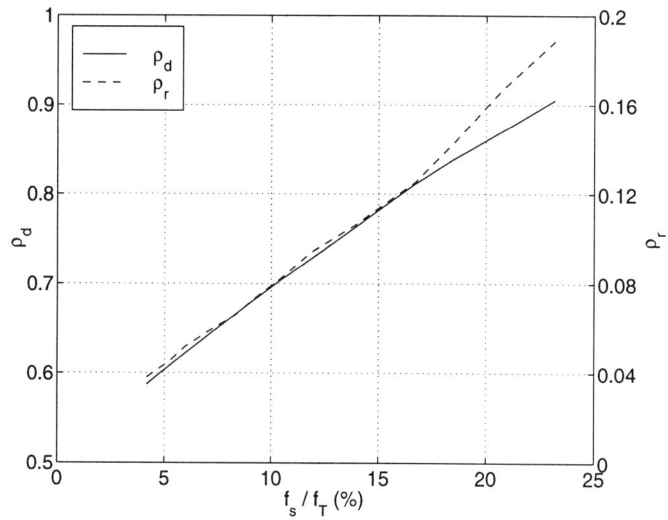

Figure 5.27a. Effect of increasing f_s on ρ_d and ρ_r.

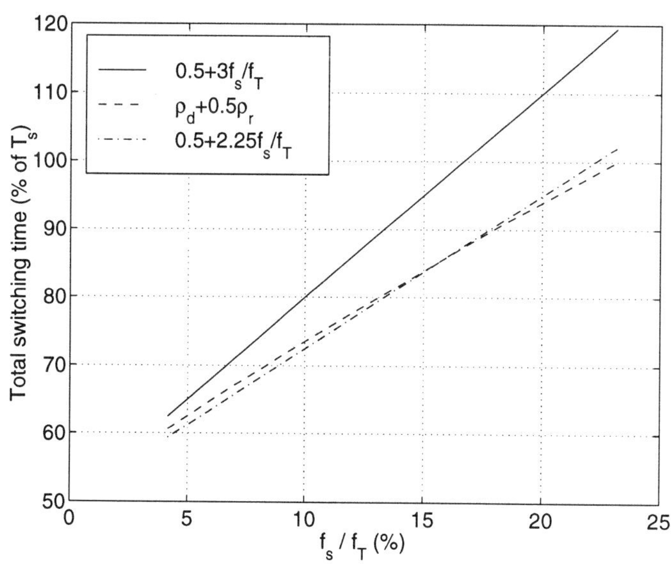

Figure 5.27b. Theoretical vs. measured switching time.

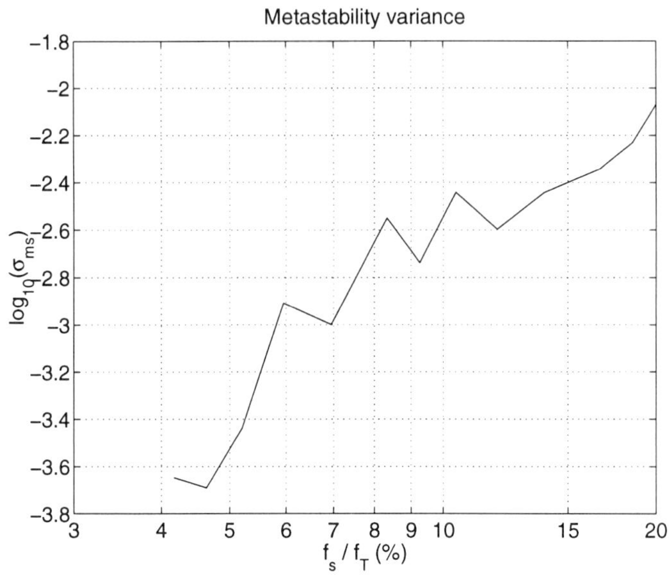

Figure 5.28a. Effect of increasing f_s on σ_{ms}.

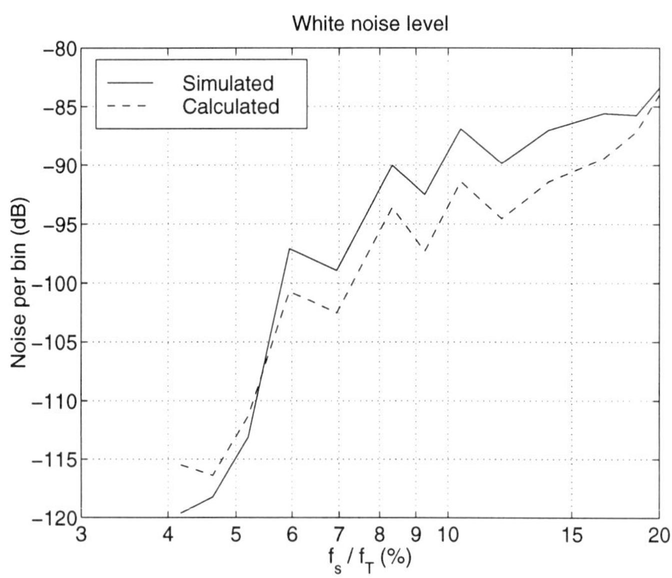

Figure 5.28b. Effect of increasing f_s on calculated and simulated noise level.

an OSR of 32 at clock frequencies from 500MHz to 2.5GHz yields the in-band white noise level per bin shown in Figure 5.28b. This agrees to within 3dB with the calculated value from (5.13) where $N = 8192$, σ_β is found from (5.24), and $\sigma_{\delta y} \approx 1.1$ is found from simulation. The modulator goes unstable at $f_s = 2.5$GHz due to excess delay and no amount of feedback tuning seems to restore stability. □

We can use the data in Figure 5.28a to come up with an approximate rule of thumb for the maximum performance achievable with a three-half latch quantizer assuming in-band noise is dominated by white noise due to metastability and $n_t = 3$ transistors in the feedback path. The calculation is shown in §5.B, and the results are equations (5.B.6) and (5.B.8):

$$\begin{array}{lcll} \text{DR} & \geq & 11.5 + 0.5\log_2 \text{OSR bits}, & f_s/f_T \leq 5\% \\ \text{DR} & \approx & 8.5 + 0.5\log_2 \text{OSR} + \log_2 \frac{f_s/f_T}{5} \text{ bits}, & f_s/f_T \geq 6\%. \end{array} \quad (5.29)$$

This tells us that clocking slower than about 5% of f_T is recommended if we desire at least 14-bit performance with a reasonable OSR like 32 or 64; better performance can be achieved with a slower clock. Clocking faster than 5% or so of f_T means we are limited to 12-bit or worse performance at the same OSRs. We do not recommend clocking faster than $f_s = 0.2f_T$ under any circumstances since stability will be questionable at best and nonexistent at worst at such high speeds.

In closing this section, we must comment further on (5.29). First, it gives an upper bound on DR: DR will be limited either by white noise due to metastability *or* quantization noise, depending on the OSR chosen. Second, it is not continuous: it has a jump between 5% and 6%. Third, the bound is not tight for $f_s/f_T < 5\%$: DR improves as we slow the clock down, though because of the semi-empirical nature of the calculations we can't easily extrapolate below this point. We estimate that metastability will have a negligible effect in most modulators when $f_s/f_T < 2\%$.

8. SUMMARY

Jitter which causes a variation in the width of the DAC pulses in a CT $\Delta\Sigma$M degrades modulator performance by whitening the in-band noise. Quantizer clock jitter and quantizer metastability are the two major mechanisms which cause this. We have derived one simple equation for each mechanism, (5.23) and (5.29), which allows estimation of maximum achievable performance. If building an integrated $\Delta\Sigma$M with an on-chip clock generated from a VCO, a properly-designed VCO should not cause a problematic level of jitter, though very high-resolution modulators or those with a center frequency away from dc might suffer. For quantizers, a three half-latch design is recommended for reducing adverse metastability effects over a simple master/slave design.

Clocking faster than about 5% of f_T will limit performance to at most 12 bits in modulators with moderate OSRs. Higher resolutions can be obtained by oversampling more or clocking more slowly.

APPENDIX 5.A: DR for VCO Clock Jitter

We derive the maximum-achievable DR for a CT $\Delta\Sigma$M clocked by a VCO with a phase noise given by (5.21). First, we start with (5.13) which is the in-band white noise level for a modulator with independent jitter and N bins:

$$10\log_{10}\left(\frac{2\sigma_{\delta y}^2 \cdot 2\sigma_\beta^2}{NT_s^2}\right). \tag{5.A.1}$$

We have omitted the -7.27dB because that is needed only for the Hann-windowed periodogram. If the in-band noise was white over the entire band, whose width expressed in bins is

$$N/(2 \cdot \text{OSR}), \tag{5.A.2}$$

then the total in-band noise would be the argument of $\log_{10}$ in (5.A.1) times (5.A.2),

$$10\log_{10}\left(\frac{\sigma_{\delta y}^2 \cdot 2\sigma_\beta^2}{\text{OSR} \cdot T_s^2}\right). \tag{5.A.3}$$

The quantity $\sigma_{\delta y}$ in (5.A.3) is found in simulation to have a value between 1 and 2, so assume

$$\sigma_{\delta y} \approx 1.5 \tag{5.A.4}$$

on average. From (5.22), and also from the second column of Table 5.2, we can find that

$$\frac{2\sigma_\beta^2}{T_s^2} = 10^{-12} f_s,\ f_s \text{ in MHz}. \tag{5.A.5}$$

Substituting (5.A.4) and (5.A.5) in (5.A.3), and recalling from Figure 5.9 in §3.2 that accumulated jitter tends to give white noise levels 1–5dB (say 3dB on average) lower than independent jitter, yields

$$10\log_{10}\left(\frac{1.5^2}{\text{OSR}} \times f_s \times 10^{-12}\right) - 3 \approx -120 + 10\log_{10} f_s/\text{OSR} \tag{5.A.6}$$

as the total in-band noise. The DR is then the maximum allowable signal amplitude minus (5.A.6); the former is given by the MSA, which for typical modulators lies between -1dB and -5dB or so. Again, assume

$$\text{MSA} \approx -3\text{dB} \tag{5.A.7}$$

on average, and note that

$$f_N \equiv \frac{f_s}{\text{OSR}}. \tag{5.A.8}$$

Using (5.A.7) and (5.A.8) with (5.A.6) gives

$$\begin{aligned} \text{DR} &\approx -3 - (-120 + 10\log_{10} f_N) \\ &= 117 - 10\log_{10} f_N \text{ dB, } f_N \text{ in MHz,} \quad (5.A.9) \\ &\approx 19 - 0.5\log_2 f_N \text{ bits, } f_N \text{ in MHz} \quad (5.A.10) \end{aligned}$$

where we have made use of (1.10) in writing (5.A.10).

APPENDIX 5.B: DR for Three Half-Latch Quantizer

Here we find the maximum-achievable DR for a CT $\Delta\Sigma$M with a three half-latch single-bit quantizer as a function of f_s/f_T. Looking at Figure 5.28b, there appear to be two distinct regions in the curve, one for $f_s/f_T < 5\%$ or so and one for $f_s/f_T > 5\%$. In the first case, the in-band noise per bin is -115dB or less in an 8192-point simulation; a single bin thus corresponds to OSR $= 4096$. If the noise were completely white, then each doubling of the OSR would raise the total noise by 3dB. Extrapolating this in the opposite direction allows us to find the total in-band noise of

$$-79 - 3\log_2 \text{OSR} \tag{5.B.1}$$

when $f_s/f_T < 5\%$. For the opposite case, the noise starts at -97dB/bin when $f_s/f_T = 6\%$ and increases roughly at 6dB/oct with f_s/f_T. Assuming white in-band noise leads to a total noise of

$$-61 - 3\log_2 \text{OSR} + 6\log_2 \frac{f_s/f_T}{5}. \tag{5.B.2}$$

DR is given by MSA minus total noise. We can see in Figure 5.26 that a modulator with half a sample of feedback delay typically has an Msa between -10dB and -6dB; assume

$$\text{MSA} \approx -8\text{dB} \tag{5.B.3}$$

on average.

Combining (5.B.3) with (5.B.1) tells us that

$$\begin{aligned} \text{DR} &\geq -8 - (-79 - 3\log_2 \text{OSR}) \quad (5.B.4) \\ &= 71 + 3\log_2 \text{OSR dB, } f_s/f_T \leq 5\% \quad (5.B.5) \\ &= 11.5 + 0.5\log_2 \text{OSR bits} \quad (5.B.6) \end{aligned}$$

where (5.B.6) makes use of (1.10). The $\geq$ sign in (5.B.4) is because the noise in (5.B.1) is worst-case, for $f_s/f_T = 5\%$; at slower clock speeds, the in-band noise will be lower and DR higher. Using (5.B.3) and (5.B.2) gives

$$\begin{aligned} \text{DR} &\approx -8 - (-61 - 3\log_2 \text{OSR} + 6\log_2 \frac{f_s/f_T}{5}) \\ &= 53 + 3\log_2 \text{OSR} - 6\log_2 \frac{f_s/f_T}{5} \text{ dB}, f_s/f_T \geq 6\% \quad (5.B.7) \\ &= 8.5 + 0.5\log_2 \text{OSR} + \log_2 \frac{f_s/f_T}{5} \text{ bits.} \quad (5.B.8) \end{aligned}$$

Once again, (1.10) was used in writing (5.B.8).

Chapter 6

A HIGH-SPEED BAND PASS $\Delta\Sigma$M DESIGN PROCEDURE

The following two chapters are devoted to the practical problem of designing a high-speed $f_s/4$ BP modulator for a radio application like that in Figure 1.7 on page 9. In Chapter 6, we outline a procedure which explains the thinking and tradeoffs that go into selecting the parameters for such a modulator, and in Chapter 7 detailed circuit simulation and test results on a fabricated SiGe part are presented.

The material that follows in this chapter was written after the modulator in Chapter 7 was designed, fabricated, and tested. At the time of its design, we did not have a rigorous design procedure for this type of modulator. In response to this, we developed the design procedure outlined here.

1. DESIGN PROBLEM STATEMENT

The question we address is, how do we choose component values in a fourth-order $f_s/4$ BP design for GHz-speed applications? This is a fairly specific problem, but we hope that by understanding its solution, insight into the design of other kinds of modulators will be gained.

A simplified single-ended model of our fabricated prototype $\Delta\Sigma$M appears in Figure 6.1. The input voltage is converted to a current through a transconductor G_{g1}, and this current drives an on-chip LC tank to produce an output voltage with a band pass shape. Transconductor G_q is connected as a negative resistor to cancel the positive resistance of the finite-Q inductor. Two second-order resonators connected in series yields an overall modulator with order four. The actual architecture is a multi-feedback one with two feedback currents per stage.

The design problem may be stated as follows: given that we desire a certain center frequency, OSR, and SNR, how do we choose the parameters L, C, R, G_g, G_q, k_2, and k_4? We make the following initial comments:

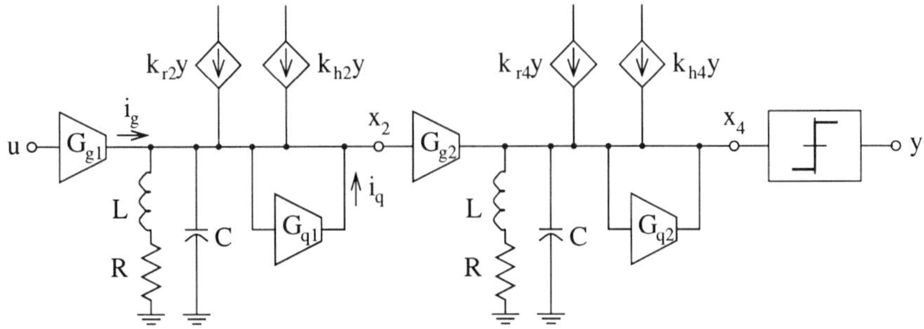

Figure 6.1. Approximate single-ended model for modulator.

- To simplify our discussion we assume resonator L and C are identical in both stages. There is no particular need for this to hold in general.
- We assume the presence of two DACs, one RZ and one HRZ (although the two specific pulse shapes aren't important), and we define

$$k_2 \equiv k_{r2} + k_{h2},\ k_4 \equiv k_{r4} + k_{h4}, \tag{6.1}$$

  i.e., the single feedback coefficients in Figure 6.1 are found by summing the amplitudes of the two DAC currents that feed back to the same place.

The selection of the modulator parameters is done by considering four main areas.

- The input transconductor (§2.) determines to a large extent the overall modulator DR. We derive expressions for the dynamic range of this transconductor in terms of its input-referred noise voltage and linearity.
- The tank components and G_q (§3.) are chosen based on the desired center frequency, inductor Q, and die area.
- The feedback DAC currents (§4.) must be selected to give the correct modulator noise-shaping characteristic and resonator node voltage scaling.
- Nonlinearity of the transconductors other than the input transconductor (§5.) leads to the folding of out-of-band noise into the signal band, and so must be controlled.

We consider each of these areas in turn.

2. INPUT TRANSCONDUCTOR G_{G1}

Many published high-speed modulators, this one included, operate predominantly in current mode. The input transconductor G_{g1} is a separate circuit

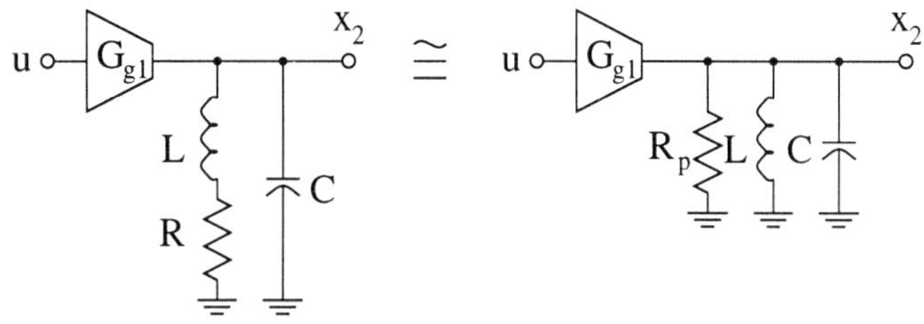

Figure 6.2. Series/parallel tank equivalence near resonance.

that is not really part of the design flow. However, overall modulator dynamic range can be no better than the DR of G_{g1} [Jen95, Mor98]: the minimum modulator input voltage u_{min} is in large part determined by the input-referred noise of G_{g1}, while the maximum voltage u_{max} is constrained by its linearity.

Obtaining a high-enough DR from G_{g1} is a circuit-design problem that is beyond the scope of what we cover here. We instead assume the existence of a transconductor circuit, and we derive approximate expressions for u_{min} and u_{max} in terms of other parameters.

Let us begin with u_{min}. We employ a series/parallel tank transformation as follows: the resistor R on the left of Figure 6.2 represents the finite Q_L of the inductor,

$$Q_L = \omega_0 L/R. \tag{6.2}$$

It can be shown [Lee98, §4.4] that over a suitably-restricted frequency range near resonance, the series LR circuit with a parallel C is approximately equivalent to a purely parallel RLC circuit as on the right of Figure 6.2 when

$$R_p = R(Q_L^2 + 1),\ L_p = L\left(\frac{Q_L^2 + 1}{Q_L^2}\right) \approx L. \tag{6.3}$$

This is useful because in a parallel RLC circuit, the impedances of L and C cancel at resonance, leaving only R_p. Next, we assume that G_{q1} is tuned such that it makes a resistance $-R_p$ to ground as depicted in Figure 6.3. We would tune G_{q1} this way because at resonance, the positive and negative R_ps cancel, which means we will have an infinite resonator Q. Ideally, this leads to an infinitely-deep notch in the quantization noise and thus good converter resolution. It now becomes possible (Figure 6.3) to write simplified expressions for the noise currents of each resistor and the feedback DAC.

Each resistor R_p and $-R_p$ has a noise current density $4kT/R_p$ A^2/Hz. If we assume the DAC is a bipolar transistor with collector current $I_c = k_2$, its noise current density term will be of the form $2qk_2$ A^2/Hz [Gra93, Chap. 11], assuming collector current shot noise dominates other noise sources such as base current and extrinsic base resistance. All these currents can be seen to drive

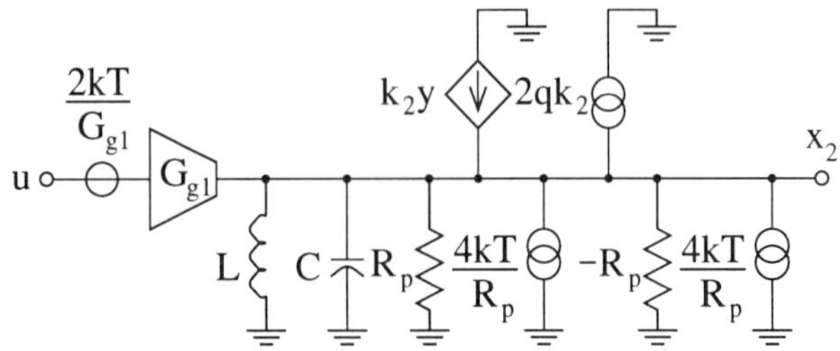

Figure 6.3. Input transconductor equivalent circuit for noise considerations.

node x_2 from ground, and therefore they may be referred to the modulator input by dividing them by G_{g1}^2, whence they become noise *voltage* densities. The input transconductor itself has a certain input-referred noise voltage density, and if we assume once again that it is a bipolar-based circuit we may write its noise voltage density as $2kT/G_{g1}$ V^2/Hz.

All the noise voltages at the input are uncorrelated, so we add them to get a total input noise voltage density of

$$\overline{v_{ng1}^2} = \frac{2kT}{G_{g1}} + \frac{2qk_2}{G_{g1}^2} + \frac{8kT\omega_0 C}{G_{g1}^2 Q_L} \ \mathrm{V}^2/\mathrm{Hz} \tag{6.4}$$

where we have used the fact that $R_p \equiv Q_L/(\omega_0 C)$ and we have assumed the noise is white in the (narrow) signal band. The total in-band noise voltage is, by definition, the minimum detectable signal

$$u_{min} \equiv \sqrt{\overline{v_{ng1}^2} \times \frac{f_s}{2 \cdot \mathrm{OSR}}}. \tag{6.5}$$

Immediately we see from (6.4) that in order to be able to detect small signals, we wish for G_{g1} to be *large*. Our first design constraint, therefore, is to make G_{g1} "large enough" to have small in-band noise. G_{g1} is proportional to current. As is often sadly the case, high DR requires high power dissipation.

The maximum input signal u_{max} is constrained by the required modulator SNR. Clearly,

$$u_{max} = u_{min} \times 10^{\mathrm{SNR}/20}, \tag{6.6}$$

which in turn determines the linearity of G_{g1}. Making the (reasonable) assumption that G_{g1} is a differential circuit with a weak cubic nonlinearity and an input-output description

$$i_{g1} = G_{g1}u - \epsilon_{g1}u^3, \tag{6.7}$$

then (6.7) can be solved to yield an input-referred third-order intercept point IIP3:

$$\text{IIP}_3 \text{ for } G_{g1} \equiv \sqrt{\frac{G_{g1}}{\epsilon_{g1}}}. \tag{6.8}$$

The required linearity for G_{g1} is then straightforward to state in terms of u_{max} and SNR: at u_{max}, we require harmonics caused by nonlinearity to be at least SNR dB below the fundamental. A simple geometrical argument [Lee98, §6.6] says that we require

$$\text{IIP}_3 \text{ for } G_{g1} = 20\log_{10} u_{max} + \text{SNR}/2 \text{ dB} \tag{6.9}$$

because the third harmonic has a slope 3 dB/dB with u while the linear term has a slope 1 dB/dB. All of this is useful when one has an actual transconductor circuit with a known dynamic range: equations (6.5), (6.6) and (6.9) can be used to see if its DR will impose an upper bound on overall modulator DR.

Although we have made a number of simplifying assumptions about the form of the transconductor circuit, it remains illustrative to write an expression for modulator SNR. We use the following fact: in order to keep the modulator stable, the feedback current must be at least as large as the maximum input current. Using (6.1), we may state

$$k_2 \geq u_{max} G_{g1} \tag{6.10}$$

(this is similar to what we found in Example 1.7). Put another way, we require

$$u_{max} \leq \frac{k_2}{G_{g1}} \tag{6.11}$$

where u_{max} is the full-scale input voltage in (6.6). Assuming we choose k_2 no larger than necessary, the inequality in (6.11) becomes an equality, and therefore the signal power is $(k_2/G_{g1})^2$. Combining this with the expression for the integrated noise power in (6.5) and simplifying leads to

$$\text{SNR}_{\text{max}} = \frac{k_2^2}{kTG_{g1}f_N(1 + \frac{k_2}{i_{g1}} + \frac{4}{Q_L}\frac{\omega_0 C}{G_{g1}})} \tag{6.12}$$

where f_N is the Nyquist bandwidth and we have used the fact that small-signal $g_m \equiv I_c/V_T$ and hence $G_{g1} = i_{g1}/(kT/q)$ for a bipolar transistor.

This expression shows that the noise is made up of the sum of three components. The one that dominates will depend on the actual design. One of the interesting insights this equation offers us is that even if G_{g1} and the DAC were noiseless, SNR would still be limited by finite Q_L. If Q_L is poor, then we need either high G_{g1} or low C to ensure that the third denominator term does not dominate; thus, poor inductor Q either increases our power dissipation or constrains our choice of capacitor size.

3. TANK COMPONENTS AND G_Q

In order to explain how to choose L, C, and the G_qs, we write an expression for x_2 in Figure 6.1. Assuming the G_{q1} transconductor draws negligible input current, we may write an equation for the Laplace transform of x_2:

$$\begin{aligned} X_2 &= [I_g + I_q + k_2 Y] Z_{eq} \\ &= [G_{g1} U + G_{q1} X_2 + k_2 Y] \frac{1}{\frac{1}{sL+R} + sC}. \end{aligned} \tag{6.13}$$

Solving for X_2 gives

$$X_2 = [G_{g1} U + k_2 Y] \frac{\frac{1}{C} s + \frac{R}{LC}}{s^2 + (\frac{R}{L} - \frac{G_{q1}}{C}) s + \frac{1}{LC}(1 - RG_{q1})}. \tag{6.14}$$

A similar equation could be written for x_4 in terms of x_2 and the second resonator parameters. The center frequency of a biquadratic transfer function is determined by the coefficient of the s^0 term in the denominator; assuming for the moment that RG_{q1} is small, we arrive at the (unsurprising) design constraint

$$\omega_0 \approx \frac{1}{\sqrt{LC}}. \tag{6.15}$$

Choosing one of L or C then fixes the other according to (6.15). Usually, the inductor series resistance R is given once L is known because one has little control over integrated inductor Q. A deep noise-shaping notch requires a high-Q resonator; a rule of thumb that ensures adequate noise-shaping for negligible performance loss from Chapter 3 §1.1 is

$$Q_{res} \approx \text{OSR}. \tag{6.16}$$

Integrated inductor Qs typically range from 5 to 10, and hence for a typical OSR of 64 we require Q-enhancement of some kind. As we have said, this is provided by the G_{q1} transconductor. A high-Q biquad has a denominator s^1 coefficient of near 0; thus, in (6.14) we need

$$\frac{R}{L} - \frac{G_{q1}}{C} \approx 0. \tag{6.17}$$

G_{q1} may now be found because it is the only unknown in (6.17), and G_{q2} may be found in a similar manner.

What are some of the considerations for how we should choose L and/or C? (6.12) shows that small C is good for noise. But smaller C means larger L, which for an integrated inductor means larger die area. As well, we can only reduce C so much before parasitics start to become significant relative to C—this could matter if C is tunable to counter process variations, and increased parasitics reduce the relative tuning range of ω_0. On the other hand, if Q_L is a function of L, then instead of choosing C it may be more sensible to choose L so that Q_L is maximized.

4. FEEDBACK DAC CURRENTS AND G_{G2}

Our choice of the feedback DAC ks and G_{g2} fixes two things: the modulator noise-shaping characteristic and the range spanned by the (stochastic) resonator voltage signals. We consider the former in this section. The latter affects the linearity of the internal modulator transconductors (i.e., those other than G_{g1}), and we treat it in the following section.

In this chapter we are discussing the design of a fourth-order $f_s/4$ BP modulator with a full sample of delay in the feedback path—recall from Chapter 5 §6.5 our recommendation to have one digital delay in a BP modulator to improve modulator immunity to metastability. This has a DT transfer function from (4.25)

$$H_{BP}(z) = z^{-1}\frac{2z^{-1} + z^{-3}}{(1 + z^{-2})^2}. \tag{6.18}$$

The equivalent CT modulator for (6.18) was found in (2.16) to be

$$\hat{H}_{BP}(s) = \frac{-1.0354(sT_s)^3 + 1.0652(sT_s)^2 - 1.3210sT_s + 4.5661}{(sT_s)^2 + (\frac{\pi}{2})^2)^2}. \tag{6.19}$$

We said in the introduction that our design employs both RZ and HRZ DACs, which as we know from Chapter 2 §1.3 gives us four individually-tunable coefficients, the DAC currents, to give us exactly the right number of degrees of freedom to control the numerator of our CT filter $\hat{H}_{BP}(s)$.

The feedback DAC levels required to implement (2.16) can be found to be

$$\begin{aligned}&(\bar{k}_{r2}, \bar{k}_{r4}, \bar{k}_{h2}, \bar{k}_{h4}) \\ = \ &(-1.1107\frac{C}{G_{g2}T_s}\frac{C}{T_s}, -0.9957\frac{C}{T_s}, +2.6815\frac{C}{G_{g2}T_s}\frac{C}{T_s}, +4.6927\frac{C}{T_s})\end{aligned} \tag{6.20}$$

(the reason for the bars over the names will be explained in the next section). C and T_s have already been determined, so $\bar{k}_{r4}$ and $\bar{k}_{h4}$ are calculable from (6.20). Moreover, the sum $\bar{k}_2 \equiv \bar{k}_{r2} + \bar{k}_{h2}$ is known from (6.10). The only remaining unknown is thus G_{g2}, which is found to be

$$G_{g2} = \frac{\pi/2}{\bar{k}_2}\frac{C^2}{T_s^2}. \tag{6.21}$$

The design is now almost complete, save the second issue above: the linearity of the internal modulator transconductors.

5. LINEARITY OF INTERNAL TRANSCONDUCTORS

When the modulator is operating, the voltages at x_2 and x_4 are stochastic in nature with approximately Gaussian distributions [Bos88]; let us denote their standard deviations σ_{x_2} and σ_{x_4}. It is not difficult to see that these values are directly proportional to the feedback current levels: driving more DAC current into x_2 and x_4 produces proportionally more voltage across the (fixed) tank impedance. If we simulate the ideal modulator in (2.16) with the DAC currents scaled as in (6.20) and zero input voltage, we find

$$(\sigma_{x_2}, \sigma_{x_4}) \approx (\frac{1}{\sqrt{2}} \frac{C}{G_{g2} T_s}, 1.157) \text{ V}. \tag{6.22}$$

$\sigma_{x_4} = 1.157$ V will in many designs overwhelm the linearity of G_{q2}, and σ_{x_2} in (6.22) will often do likewise to G_{q1} and G_{g2} for practical parameter values. It is thus preferable to scale all DAC currents by a parameter γ_k:

$$(k_{r2}, k_{r4}, k_{h2}, k_{h4}) = (\gamma_k \bar{k}_{r2}, \gamma_k \bar{k}_{r4}, \gamma_k \bar{k}_{h2}, \gamma_k \bar{k}_{h4}) \tag{6.23}$$

and

$$G_{g2} = \frac{\pi/2}{k_2} \frac{C^2}{T_s^2} \gamma_k \tag{6.24}$$

where k_2 is from (6.10). This merely scales the transfer function in (2.16), which does not alter its noise-shaping characteristics, while simultaneously scaling resonator voltages:

$$(\sigma_{x_2}, \sigma_{x_4}) \approx (\frac{1}{\sqrt{2}} \frac{C}{G_{g2} T_s} \gamma_k, 1.157 \gamma_k). \tag{6.25}$$

Therefore, γ_k lets us reduce voltage swings so as to meet any linearity requirements on G_{q1}, G_{g2}, and G_{q2}. The overbars in (6.20) are thus to denote nominal feedback levels; actual levels are found from (6.23).

Suppose these three transconductors have known IIP_3 with a form similar to (6.8). We normally think of linearity in terms of the harmonics produced in response to multiple sinusoidal inputs. Such thinking is not helpful here because the internal transconductors are mainly driven by the modulator output (a wideband stochastic signal) rather than a sinusoid. Understanding the effect of nonlinearity is aided by writing out the time-domain differential equations (DEs) for the state variables.

For simplicity, assume G_{q1} is the only nonlinear transconductor and that $R = 0$. A second-order differential equation for x_2 can be written as two coupled first-order equations by introducing an auxiliary state variable x_1:

$$\frac{dx_1}{dt} = \frac{1}{LC} x_2, \quad \frac{dx_2}{dt} = -x_1 + \frac{1}{C} \left[G_{g1} u + \epsilon_{q1} x_2^3 + k_2 y \right]. \tag{6.26}$$

We divide the discussion into two cases.

1. Suppose first that $\epsilon_{q1} = 0$ (i.e., that G_{q1} is linear). The term inside square brackets in (6.26) for dx_2/dt then reduces to $[G_{g1} u + k_2 y]$. This term

describes ideal modulator operation: the input u is combined with the fedback output bit y, *and* the amplitude of u in the spectrum of y is determined from k_2/G_{g1} as in (6.11).

2. Now suppose there is no input ($u = 0$) and $\epsilon_{q1} \neq 0$ (G_{q1} is nonlinear). The square-bracketed term becomes $[\epsilon_{q1}x_2^3 + k_2 y]$. By analogy to the first case, the input u has been replaced with x_2^3; therefore, by analogy, we expect the spectrum of x_2^3 to appear in the spectrum of y with an amplitude related to the ratio k_2/ϵ_{q1}.

From theory, the spectrum of $x_2(t)^3$ is $X_2(s)*X_2(s)*X_2(s)$, where the asterisk denotes convolution. Simulated spectra of y, x_2, and x_2^3 appear in Figures 6.4a and 6.4b. We observe that x_2 has a noise-shaped appearance similar, but not identical, to that of the output bit stream y. The convolution causes folding of some of the out-of-band noise into the signal band, but how much?

We can answer that question with the help of normalization. If we simulate the modulator and plot its DR and peak SNR, SNR_{max}, as a function of normalized ϵ_{q1},

$$\bar{\epsilon}_{q1} \equiv \frac{\epsilon_{q1}\sigma_{x_2}^3}{k_{r2} + k_{h2}}, \tag{6.27}$$

the graphs appear as in Figure 6.5a and look the same for a given OSR. The form of $\bar{\epsilon}_{q1}$ is logical given (6.26) and point 2 above. Similar normalizations can be found for the other two nonlinear parameters:

$$\bar{\epsilon}_{g2} \equiv \frac{\epsilon_{g2}\sigma_{x_2}^2}{G_{g2}}, \tag{6.28}$$

$$\bar{\epsilon}_{q2} \equiv \frac{\epsilon_{q2}\sigma_{x_4}^3}{k_{r4} + k_{h4}}. \tag{6.29}$$

Graphs of DR and SNR_{max} are plotted in Figures 6.5b and 6.5c. From these graphs, we may derive the following rules of thumb for the restrictions on the $\bar{\epsilon}$s for each internal transconductor that will not affect DR significantly:

$$\bar{\epsilon}_{q1} < 10^{-3},\ \bar{\epsilon}_{g2} < 10^{-2},\ 10\log_{10}\bar{\epsilon}_{q2} < -5 - \log_2 \text{OSR}. \tag{6.30}$$

The final portion of the design procedure can now be described. The nominal feedback $\bar{k}$s and signal levels σ_xs are found from the other known parameters by using (6.24) for G_{g2} followed by (6.23) and (6.25), where to start we assume $\gamma_k = 1$. The IIP_3s for the internal transconductors are characterized, and the normalized $\bar{\epsilon}$s are calculated using (6.27)–(6.29). If they are large enough to violate (6.30), then γ_k can be lowered, which will lower the $\bar{\epsilon}$s by a factor of γ_k^2.

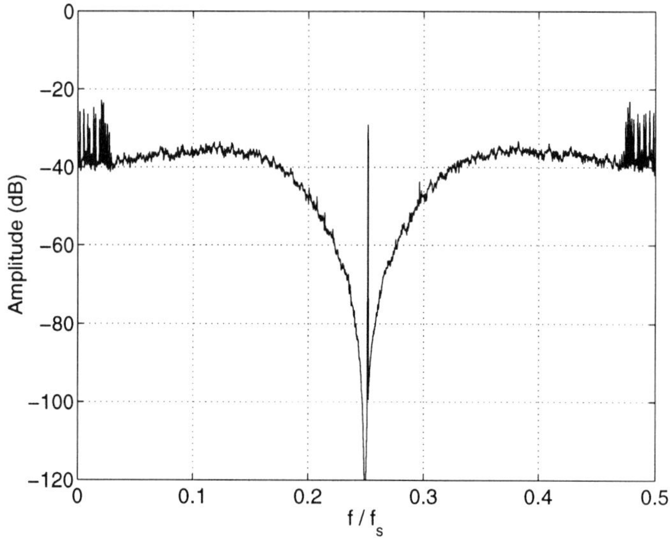

Figure 6.4a. Spectrum of modulator output bit stream y.

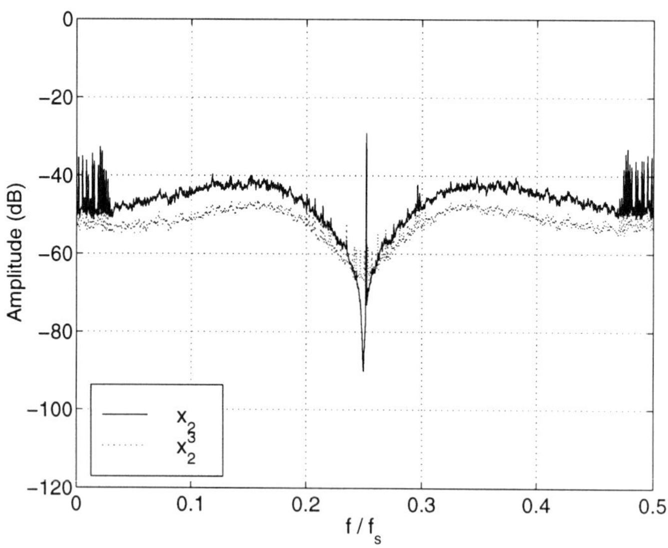

Figure 6.4b. Spectrum of first resonator output voltage x_2, and x_2^3.

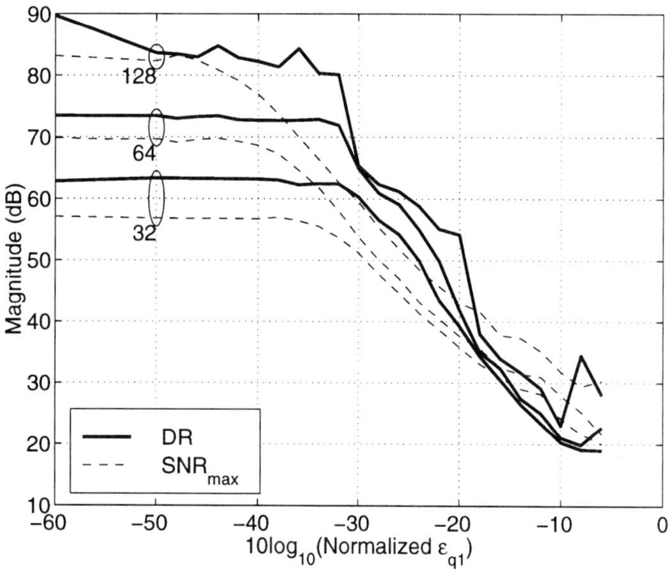

Figure 6.5a. DR and SNRmax for $\bar{\epsilon}_{q1}$. Numbers on curves are OSR values.

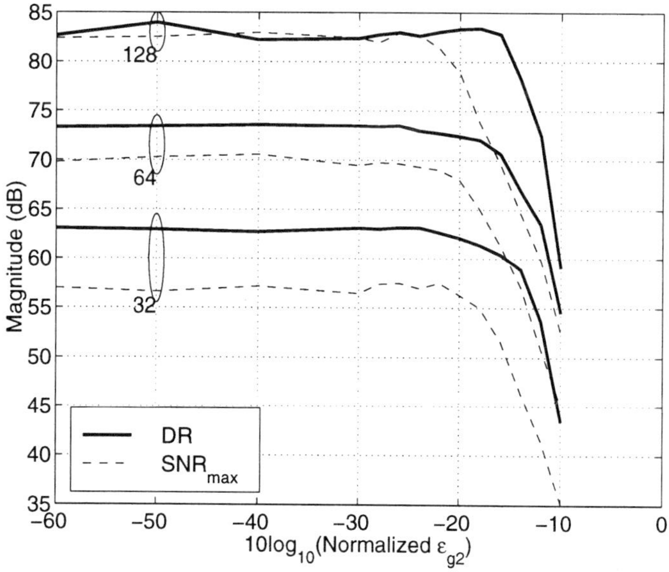

Figure 6.5b. DR and SNRmax for $\bar{\epsilon}_{g2}$. Numbers on curves are OSR values.

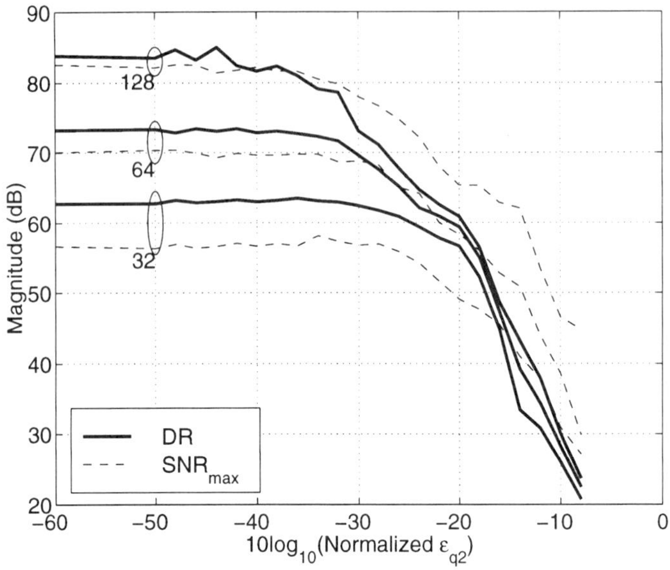

Figure 6.5c. DR and SNR$_{max}$ for $\bar{\epsilon}_{q2}$. Numbers on curves are OSR values.

We should also keep in mind that we can also alter the signal levels σ_{x_2} and σ_{x_4} by altering the tank impedance, if for some reason changing γ_k is found to be unsatisfactory. This will require iteration in the procedure, but such iteration will likely be required in other parts of the procedure anyway.

6. SUMMARY

We summarize the salient points of our design method here. Take as given center the frequency f_0 ($f_0 = \omega_0/2\pi = f_s/4$), SNR (assumed for simplicity equal to DR), and conversion bandwidth $f_N/2 = f_s/(2 \cdot \text{OSR})$. For the input transconductor:

- Design the circuit and find the achieved G_{g1} and ϵ_{g1} in (6.7)
- Use SPICE to find the total input-referred noise voltage over the Nyquist band
- Calculate u_{min} from (6.5), u_{max} from (6.6)
- Ensure IIP_3 satisfies (6.9)

For the tank parameters:

- Calculate the first feedback current k_2 from (6.10)

- Calculate the LC product from (6.15)
- Determine the inductor series resistance
- Calculate the required G_{q1} and G_{q2} from (6.17)
- With DAC and G_{q1} circuits present, resimulate input-referred noise in SPICE and ensure the input transconductor still has the necessary dynamic range

For the feedback DAC levels and internal transconductors:

- Calculate the required G_{g2} from (6.24)
- Calculate the nominal $\bar{k}$s and σ_xs from (6.23) and (6.25) using $\gamma_k = 1$
- Design the other transconductors to meet G_q and G_{g2} specs and find the achieved ϵ_q and ϵ_{g2}
- Calculate the normalized $\bar{\epsilon}$s from (6.27)–(6.29)
- Check if performance loss is significant with (6.30) and adjust γ_k and/or tank impedances

As we have said, application of this procedure will involve a good deal of iteration. Moreover, we recommend SPICE or some other full-circuit simulator for noise measurements; our estimates in (6.4) and (6.12) are only very approximate.

Chapter 7

A 4GHZ FOURTH-ORDER BAND PASS ΔΣM

The final major chapter of this monograph deals with the detailed circuit design, test, and redesign of a BP modulator which clocks at $f_s = 4$ GHz for conversion of 1 GHz IF signals. The circuit was fabricated in a 0.5μm SiGe BiCMOS process, though it is an all-bipolar design with HBTs rated at $f_T = 50$GHz and $f_{max} = 60$GHz [Har95]. Preliminary circuit details and test results were given in [Gao98a]; here, we give the circuit a much more detailed treatment.

A block diagram of the circuit appears in Figure 7.1. It looks very similar to the single-ended block diagram of Figure 6.1, and is also reminiscent of Figure 4.1 in style, only with resonators instead of integrators. An input voltage is fed through a transconductor G_g which produces a current $i_g = G_g v_i$ to drive an on-chip parallel LC tank. This gives the tank output voltage v_o a band pass shape:

$$V_o = I_G Z_{LC} = \frac{G_g V_i}{sC + 1/sL} = \frac{(G_g/C)s}{s^2 + 1/LC} V_i. \tag{7.1}$$

The series connection of two second-order resonators yields a fourth-order modulator. As in Chapter 6, the integrated inductor has a poor quality factor Q_L, so a Q-enhancement transconductor G_q is connected as a negative resistor to cancel the positive resistance of the inductor. Both the gain A_0 and the Q of the resonator Q_{res} are tunable. The quantizer and latches are such that this is a one digital delay multi-feedback architecture, and the feedback operates via current-summing with simple tunable current-switching DACs.

1. PARAMETERS FOR THIS DESIGN

As noted in the introduction to Chapter 6, this circuit was designed long before the procedure in that chapter was formalized. It is nonetheless instructive

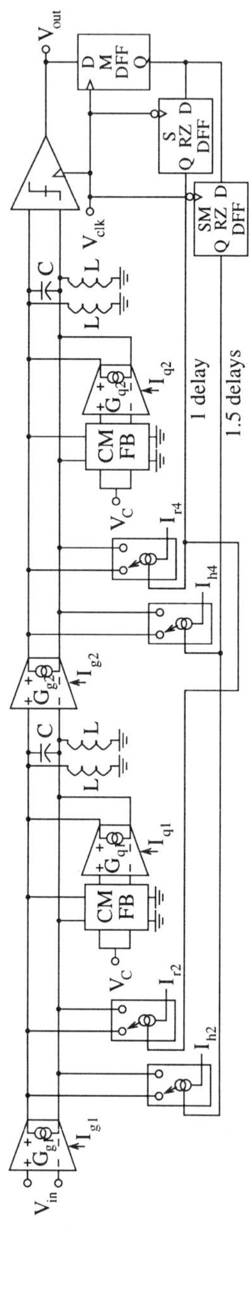

Figure 7.1. A 4 GHz fourth-order BP CT ΔΣM.

Table 7.1. Parameters for the fourth-order design in this chapter.

Tank element values		$L = 3.5$nH
		$C = 6.1$pF
		$R = 2.72\Omega$
G_g	Range	$2 \rightarrow 8$ mA/V
ϵ_g	Typical	9×10^{-3} mA/V^3
G_q	Range	$2 \rightarrow 9$ mA/V
	For $Q = \infty$, calc	4.3mA/V
	For $Q = \infty$, meas	8.9mA/V
ϵ_q	Typical	5×10^{-4} mA/V^3
k_r, k_h	Range	$0 \rightarrow 500$ μA
σ_{x2}, σ_{x4}	Range	$4 \rightarrow 12$ mV
$\overline{v_{ng1}}$	Calc (typical)	3.5nV/$\sqrt{\text{Hz}}$
	Sim (typical)	20nV/$\sqrt{\text{Hz}}$
u_{min}	Typical	90μV
G_{g1} IIP$_3$	Sim	-2.3dBV
$\bar{\epsilon}_{q1}$	Maximum	$2.3 \times 10^{-6} = -56.4$dB
$\bar{\epsilon}_{g2}$	Maximum	$1.8 \times 10^{-4} = -38.5$dB
$\bar{\epsilon}_{q2}$	Maximum	$2.6 \times 10^{-6} = -55.8$dB
SNR limit	u_{min}, IIP$_3$	52dB

to calculate its parameters as defined there. We have done this for an input signal bandwidth of 20 MHz in Table 7.1, which corresponds to an OSR of 100. Worthy of mention are the following points:

1. SPICE measures typical in-band noise voltage densities of 20 nV/$\sqrt{\text{Hz}}$, but our approximate formula (6.4) is too optimistic: it predicts 3.5 nV/$\sqrt{\text{Hz}}$. Our formula was based on one transistor and no base-resistance noise—neither of which is valid in our circuit. Tests indicate both SPICE and (6.4) predict the same dominant term: the noise seems to come mostly from the DAC, the middle term in (6.4). As well, SPICE validates the assumption that the noise is white across the signal band.

2. The swings at the resonator outputs σ_{x_2} and σ_{x_4} can be made only as large as about 12 mV because the DAC currents were chosen very small in this design. As a result, when we find the normalized $\bar{\epsilon}$ values and compare them to (6.30), it happens that none of them are large enough to affect DR appreciably at OSR $= 100$.

3. The 20 nV/$\sqrt{\text{Hz}}$ input-referred noise for G_{g1} in a 20 MHz bandwidth gives a minimum-detectable input signal of $u_{min} = 90$ μV. The linearity of that transconductor is such that IIP$_3 = -2$ dBV or so, and calculation with (6.6) and (6.8) gives a maximum SNR of 52 dB.

Thus, we predict at the outset that measurements will show this to be no better than about an eight-bit converter.

2. CIRCUIT BLOCKS

In this section we cover the design of each of the major circuit blocks: transconductors, resonators, latches, and DACs[1].

2.1 RESONATOR

Before proceeding to the final resonator circuit, it is useful to break it down into separate components. It turns out that the model in Figure 6.1 approximates the resonator quite well. A lumped-element equivalent circuit for each on-chip inductor half-circuit can be derived from the physical inductor layout which includes metal resistance, inter-turn capacitance, capacitance through the dielectric to the substrate, etc. A SPICE ac analysis of the lumped equivalent produces the characteristics in Figures 7.2a and 7.2b. The inductor's nominal value at $f_s/4 = 1$GHz is $L = 3.5$nH with a Q of about 8.1, and its self-resonant frequency is about 12.8GHz. This means the series resistance is about $R = 2.72\Omega$; it is mildly frequency-dependent, but in simulation the dependence is weak enough that we may consider R to be constant. C is actually two 1.525pF capacitors in parallel with each other with both ends connected to either collector of a differential pair. Thus, the equivalent capacitance to ac ground is $C = 6.1$pF, four times this value.

TRANSCONDUCTOR

Both input and Q-enhancement transconductors are so-called *multi-tanh* circuits [Gil98] whose operation can be understood starting with the simple differential pair in Figure 7.3. A differential input voltage v_{in} becomes a differential output current i_{out} with a hyperbolic tangent characteristic [Gra93, Chap. 3]; the transconductance $G_m \equiv di_{out}/dv_{in}$ thus has a sech^2 shape. Figure 7.4a shows how the G_m vs. v_{in} curve varies as a function of tail current I_{tail} for $A = 1$: we can increase the peak G_m by increasing I_{tail}. The input range over which the circuit is linear can be improved by first unbalancing the differential pair, where A transistors are connected in parallel on one side (effectively creating a transistor whose emitter is A times larger). This has the effect of shifting the peak of the G_m vs. v_{in} curves as illustrated in Figure 7.4b. Next, unbalanced pairs have their outputs cross-coupled, Figure 7.5, which results in the overall G_m characteristic having a double-hump shape, Figure 7.6a. The

[1]Until now we have specified input levels in dB, which as we noted in Chapter 2 is dB relative to full scale. In this chapter, because this is a real circuit with an input voltage, we talk about input levels both in V and (because the circuit is intended for a radio receiver) dBm assuming a 50Ω impedance. The one place we refer to dB, Figure 7.38, still uses dB relative to full scale, and we calculate the full-scale input level in §3.1.

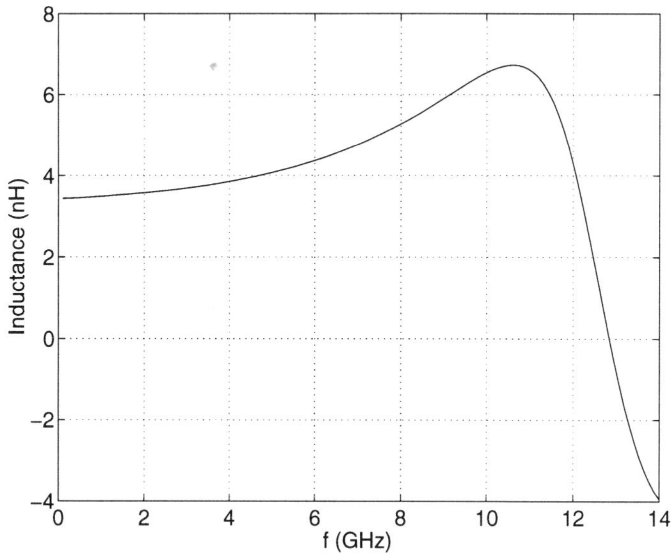

Figure 7.2a. SPICE ac analysis of inductor L.

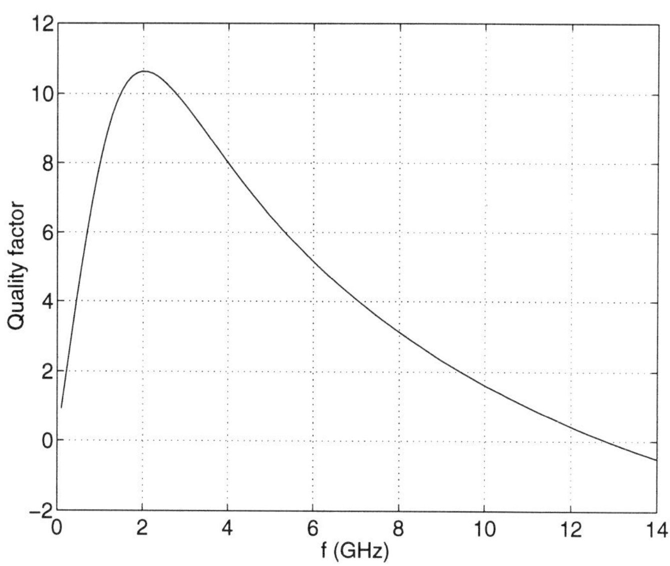

Figure 7.2b. SPICE ac analysis of inductor Q.

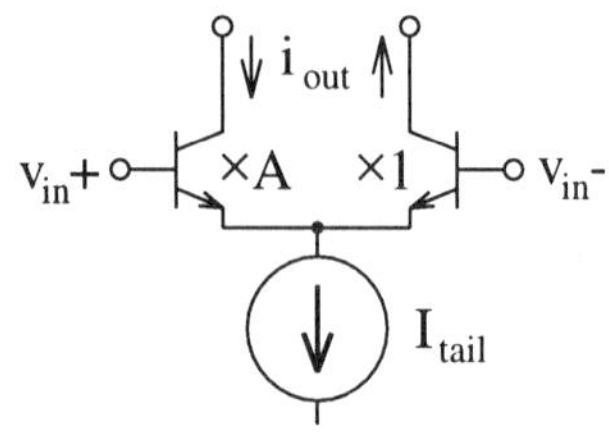

Figure 7.3. Differential pair transconductor.

horizontal shift of the G_m curves is altered by changing A, and we also have the option of adding small emitter resistors R_e to further change the shape of the individual pairs' G_m curves, Figure 7.6b. By correctly choosing A and R_e, we can get a flat top on the final G_m vs. v_{in} curve. This is what gives us the desired increase in linear range.

The main advantage of using a multi-tanh circuit for linearity over a differential pair with emitter degeneration is that the latter has a fixed transconductance, while in the former G_m is tunable with I_{tail} while retaining good linearity. Furthermore, the increase in noise suffered by using two pairs of transistors and two tail current generators (instead of one of each in a simple differential pair) is more than made up for by the increase in 1dB compression point—multi-tanh circuits have higher DR than degenerated differential pairs when both noise and linearity are taken into account.

INPUT TRANSCONDUCTOR

The actual multi-tanh topology used in this architecture is depicted in Figure 7.7. In place of resistors, diode-connected transistors are used. Between the input and ac ground, we have $D = 3$ diodes formed by base-emitter junctions; what are the ramifications of employing this configuration over $D = 1$ as in the original multi-tanh design in Figure 7.5 or $D = 2$?

Example 7.1: Let us study the circuit in Figure 7.7 with 500Ω load resistors on the collectors of the input transistors going to $V_{CC} = 5$V. We will use an input common-mode (CM) level of 3.4V, which is about what the actual level in the final modulator is.

For $I_{tail} = 0.4$mA, Figure 7.8a plots G_g vs. V_{in} from a SPICE dc analysis for $D = 1$, 2, and 3 base-emitter diodes. G_g falls from 9.40mA/V to 4.70mA/V and 3.68mA/V: it is inversely proportional to D. At the same time, the linear range in Figure 7.8b increases proportionally to D: a 1% drop in G_g happens at $V_{in} = 25$mV, 50mV, and 75mV. This behavior is easily understood by considering the series connection of D identical diodes with the same current I and voltage V across their terminals, Figure 7.9.

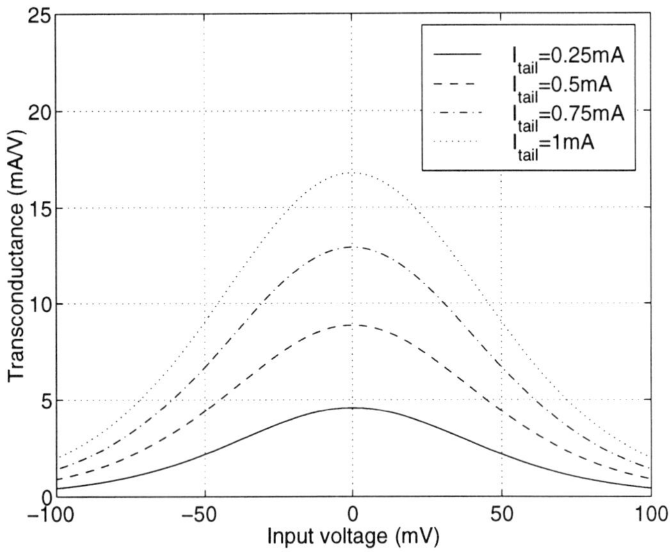

Figure 7.4a. Transconductance of differential pair as a function of tail current.

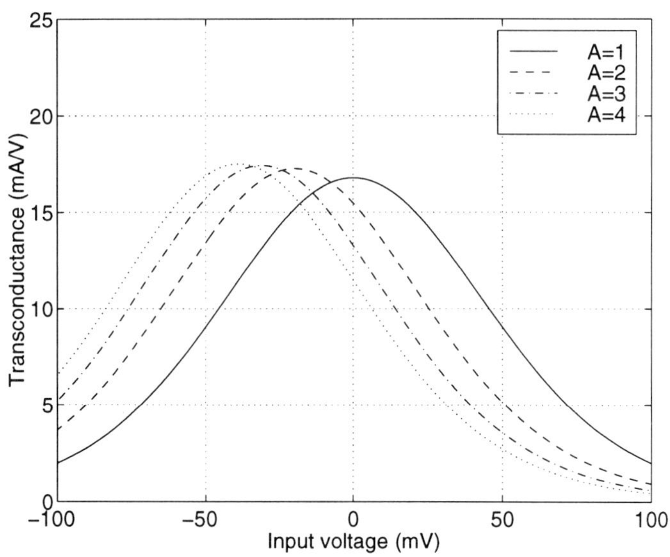

Figure 7.4b. Transconductance of differential pair as a function of number of transistors.

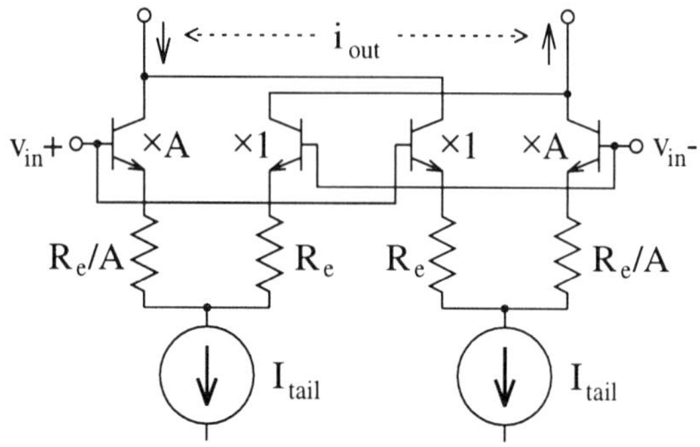

Figure 7.5. Multi-tanh circuit.

The current through each diode is the same and is given by

$$I = I_s \exp\left(\frac{V/D}{kT/q}\right) \tag{7.2}$$

because the voltage across each diode is V/D. The small-signal transconductance is found from

$$g_m = \frac{dI}{dV} = \frac{I_s}{D(kT/Q)} \exp\left(\frac{V/D}{kT/q}\right) = \frac{1}{D}\frac{I}{kT/q}. \tag{7.3}$$

The transconductance G_g above falls because of the $1/D$ factor in front of (7.3), and the linear range increase arises because of the V/D inside the exponential.

The advantage of using diodes rather than passive resistors should be clear: the g_m for a diode is proportional to I_{tail}, but for a passive resistor g_m is fixed. This explains why in Figures 7.10a and 7.10b, which contrast the two cases for similar G_g and I_{tail}, the linearity is retained as G_g is varied in the diode case but not in the resistor case.

A fair comparison of the D choices includes several parameters: the realizable range of G_g values, I_{tail} per G_g (which gives a measure of power dissipation), and the dynamic range, which is a combination of the noise figure and the linearity. We will use an input frequency of 1GHz, since that is approximately the frequency at which the circuit must operate in the ΔΣM. The parameters for this cell are presented in Table 7.2. A discussion of the results is in order.

- The realizable G_g range is unlimited in theory as long as we are willing to supply the current I_{tail}. What limits us in this design is headroom, in particular the design of the biasing circuit that supplies I_{tail}. For $D =$

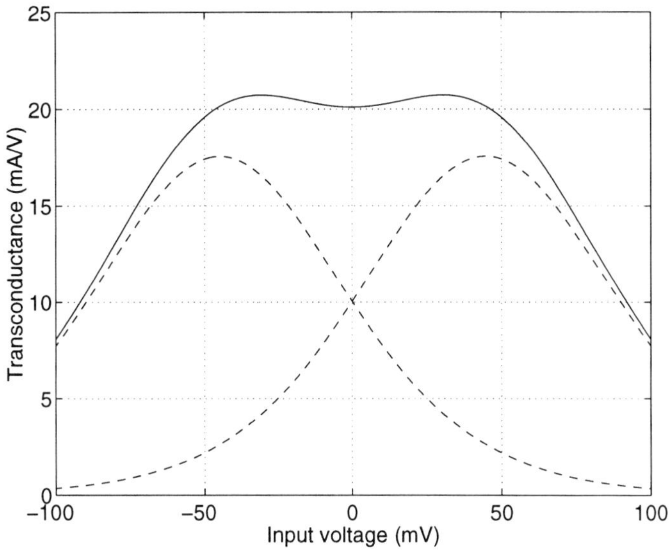

Figure 7.6a. Multi-tanh G_m characteristic.

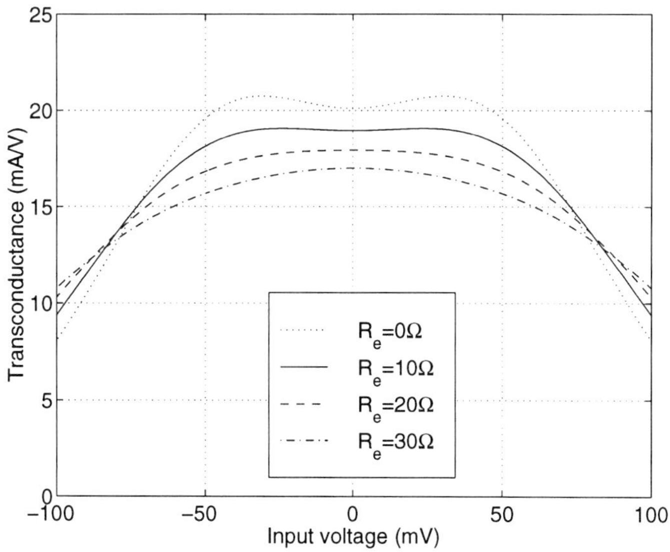

Figure 7.6b. Effect of varying R_e.

Table 7.2. Comparison of multi-tanh architectures.

Parameter		$D = 1$	$D = 2$	$D = 3$
Realizable G_g range (mA/V)		2–22	2–11	2–8
ΔI_{tail} (μA per mA/V)		45	90	135
$\mathrm{NF_{opt}}$ (dB), $R_{S\mathrm{opt}}$ (Ω)	$I_{tail} = 400\mu$A	5.67, 600	7.51, 900	9.11, 1100
	$I_{tail} = 1$mA	4.73, 400	5.83, 550	6.86, 700
Linearity	1dB compression (dBm)	−11.5	−5.5	−1.9
	Estimated $\mathrm{IIP_3}$ (dBm)	−1.9	+4.1	+7.7
Approximate ΔDR (dB)		0	5	7

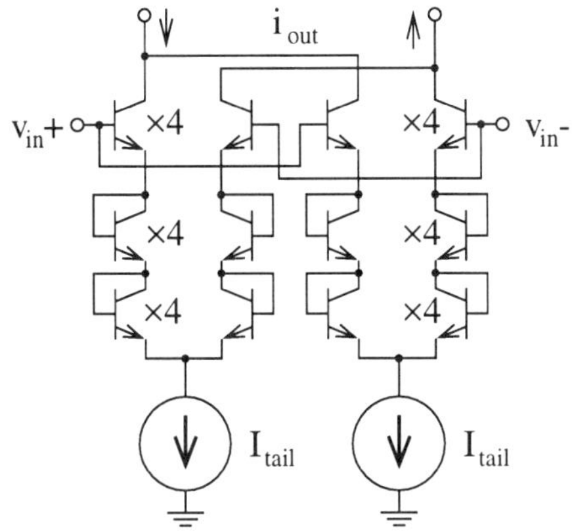

Figure 7.7. Actual transconductance topology used.

3, excessive I_{tail} pushes the transistor supplying I_{tail} into saturation which degrades linearity. This could be remedied with increased supply voltage V_{CC}, though this would increase V_{CE} on all the transistors and possibly introduce problems with BV_{CEO}.

- As expected, power dissipation is proportional to D for a given G_g. Again, high G_g can be obtained no matter what D is as long as we are prepared to dissipate more power.

- Noise figures were measured in SPICE at 1GHz using ac analysis; the source resistance was swept until the optimum NF was found, and both values are listed. NF falls as I_{tail} increases; the NF_{opt} increases between 1 and 1.5dB with D depending on I_{tail}. As well, larger D requires higher $R_{S_{opt}}$ for optimum NF. That being said, NF and $R_{S_{opt}}$ are roughly constant for the same G_g.

- Linearity was measured with transient simulation in SPICE. The 1dB compression point is fairly easy to measure by sweeping the input voltage, plotting the output current, and using straight-line extrapolation to find where it deviates from linear by 1dB. Figure 7.11 shows the results of a two-tone test in SPICE where the input tones are at 980MHz and 1GHz, and the third harmonic at 1.02GHz is plotted. Unfortunately, the third harmonic does not behave in a Taylor-series manner: the slope of its magnitude doesn't increase by 3dB per dB of input voltage, as the dashed and dotted lines show. This is an inherent property of multi-tanh circuits [Gil98]; because of it, defining IIP_3 is difficult. The definition we adopt (simply so we have a method of discussing it) is that it is 9.6dB higher than the 1dB compression point, which derives from the

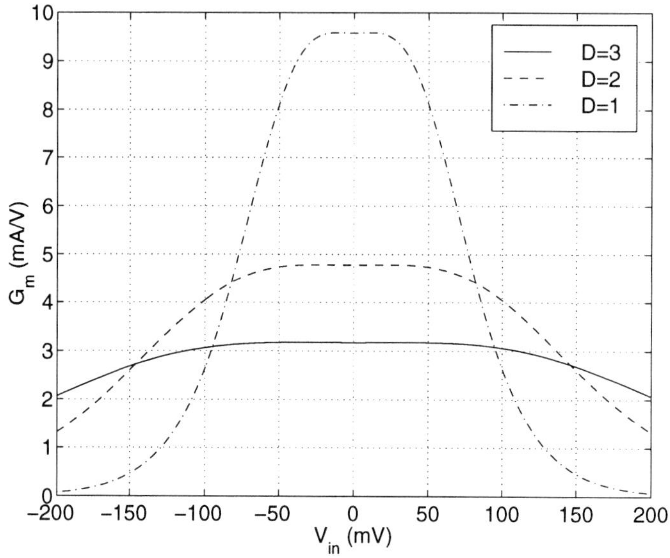

Figure 7.8a. Absolute transconductance as a function of number of diodes.

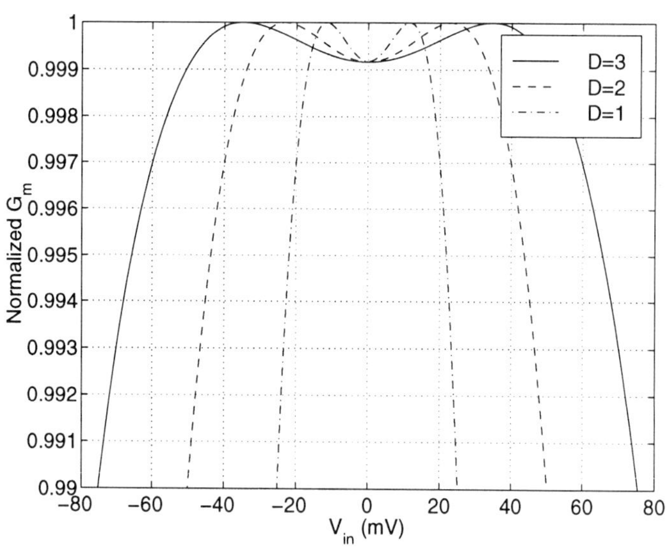

Figure 7.8b. Transconductance normalized to peak G_g as a function of number of diodes.

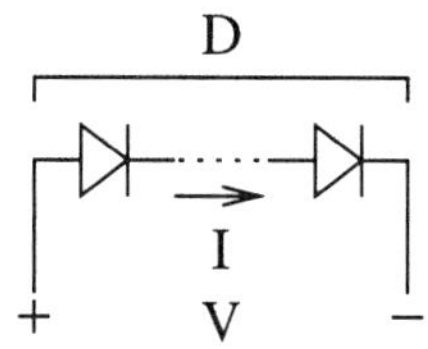

Figure 7.9. Series-connected diodes.

assumption of a weak Taylor-type cubic nonlinearity in (6.7) and (6.8). In any case, linearity improves roughly as $20 \log_{10} D$.

- Combining the previous two facts—the small increase in NF with the larger increase in 1dB compression point—leads to the final table row which shows DR improvement resulting from increasing D.

In conclusion, there is a clear DR advantage of using $D = 2$ instead of $D = 1$, and a slight further advantage in using $D = 3$ over $D = 2$. In a $V_{CC} = 5$V design such as this one, $D = 3$ is about the maximum we can get away with while retaining acceptable headroom. □

Q-ENHANCEMENT TRANSCONDUCTOR

The band pass resonator block appears in Figure 7.12. The input voltage is applied through emitter followers (not shown) to the input multi-tanh block whose load is the LC tank. The Q-enhancement transconductor is a second multi-tanh circuit whose input is derived from the output voltage sampled by a capacitive divider. The dc level at the inputs to the Q-enhancement multi-tanh would not be well-defined without the control voltage V_{CDB} and the circuitry associated with it.

Examination of the dc operating point of the circuit makes it clear that care must be taken when choosing where to set various voltages when the power supply is 5V. The bias circuit has to be turned off in practice because the base currents of the Q multi-tanh input transistors alone drop enough voltage across the 2kΩ and 10kΩ resistors to push the input common-mode voltage V_{IB} to 3.5V, which, as was also true with the input transconductor, strains the ability of the transistors supplying I_{tail} to stay out of the saturation region. Raising V_{CDB} to a voltage which turns Q_{CDB} on, pulls V_{IB} down further, resulting in worse IIP_3 for the G_q transconductor. Therefore, $V_{CDB} = 0$V is required in normal circuit operation.

The G_q transconductor's ac input voltage derives from V_{out} through the 1pF/3pF capacitive voltage divider. We expect a voltage division of 0.25 (i.e., the input signal at V_{IB} smaller than V_{out} by a factor of four), though SPICE simulations indicate a value of closer to 0.20 for a signal between 1GHz and

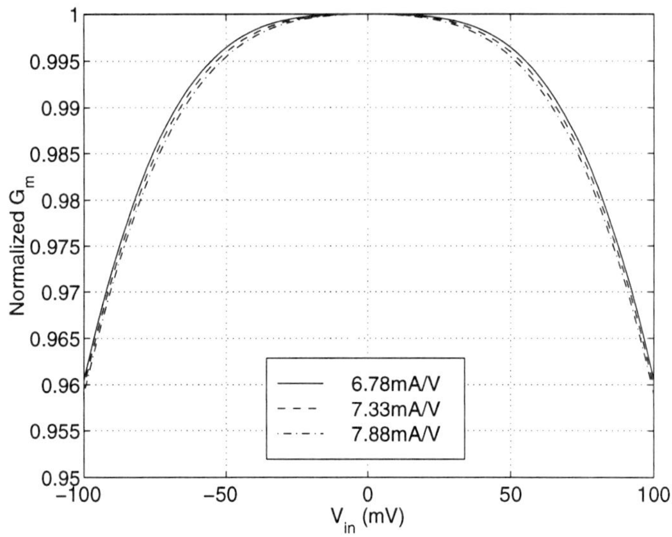

Figure 7.10a. Linearity for emitter diodes vs. emitter resistors ($A = 4$, $D = 3$).

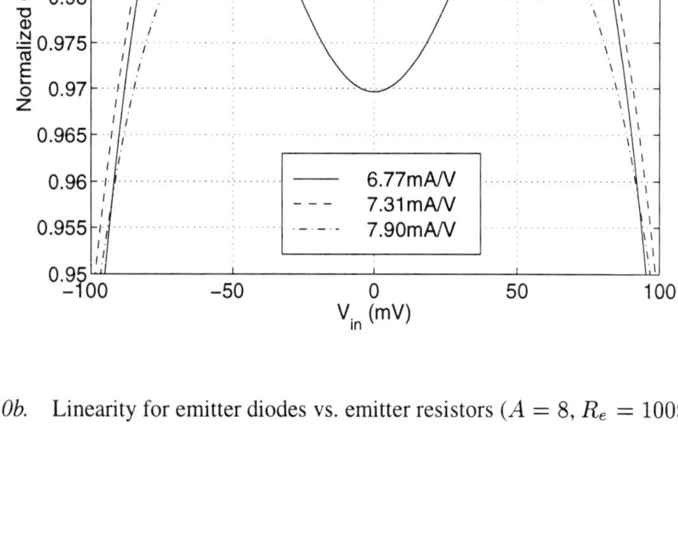

Figure 7.10b. Linearity for emitter diodes vs. emitter resistors ($A = 8$, $R_e = 100\Omega$).

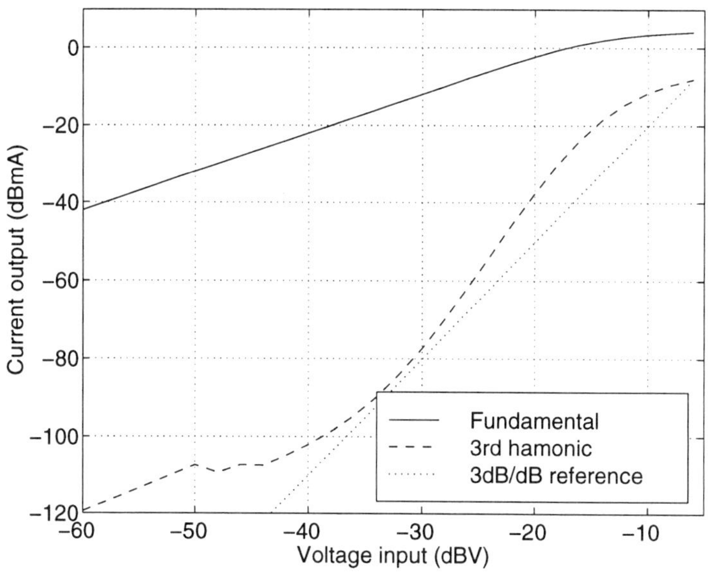

Figure 7.11. Linearity plot for a multi-tanh circuit.

4GHz. The transistors in the G_q multi-tanh circuit are, however, four times larger than those in the G_g circuit, so they are capable of supplying about four times the current at the same V_{BE}. The net result is that the achievable transconductance range obtainable for the G_q transconductor is not markedly different from that for the G_g transconductor. The linearity is better, however: G_q achieves IIP$_3$ $= +20$dBm or so vs. $+8$dBm for G_g, though this is expected because of the voltage division.

RESONATOR CHARACTERISTICS

We illustrate gain and Q tuning for the overall resonator in a SPICE ac analysis. Figure 7.13a shows V_G tuning with fixed V_Q, where these two voltages control the tail currents in each multi-tanh block. The peak gain varies over about 9.5dB, with Q_{res} remaining almost constant at about 73; the peak gain is proportional to the G_g in the gain multi-tanh circuit. Figure 7.13b is with V_G fixed and V_Q tuned, and the Q_{res} varies from 28 for $V_Q = 2.4$V to about 360 for $V_Q = 2.7$V. These results are useful because they let us estimate actual voltage levels which result in a certain gain and Q during testing.

We also do a comparison of ac analysis to transient analysis. For $V_G = 2.2$V and $V_Q = 2.6$V, analyses of gain in Figure 7.14a and phase in Figure 7.14b are contrasted. It takes the output amplitude about 400 cycles to settle, so

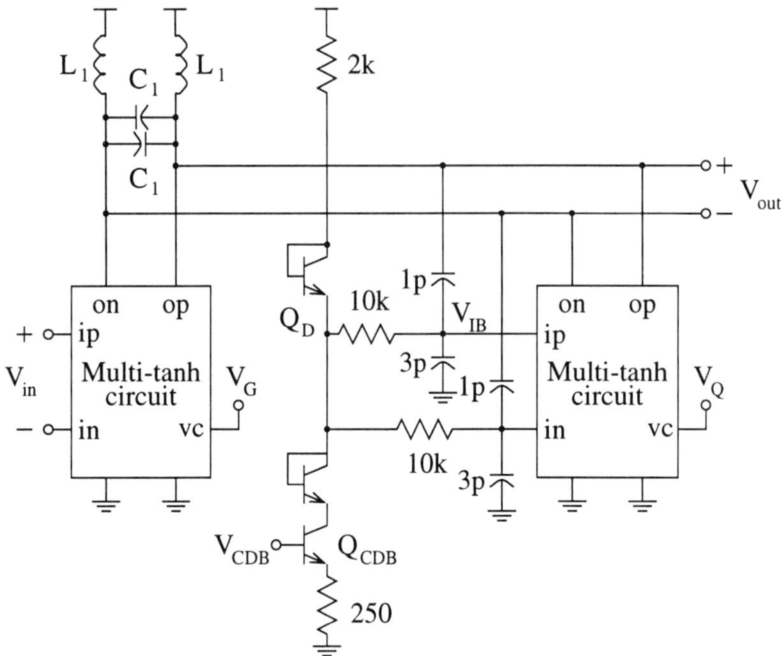

Figure 7.12. Band pass resonator block.

each simulation takes quite a bit of time. Agreement is good except near the peak gain where it becomes merely acceptable: ac analysis predicts $Q = 74$ while the transient analysis shows $Q \approx 150$. SPICE ac analysis is linearized while transient analysis preserves nonlinearities, hence the two disagree here [Che94, Che98c].

We can find the input-referred noise (and hence the minimum-detectable input signal u_{min}) from a SPICE ac simulation. A value for typical control voltage settings is $\overline{v_{ng1}} = 20\text{nV}/\sqrt{\text{Hz}}$, though this can vary by a factor of two either way depending on the exact biasing. In a 20MHz band, the total noise comes out to about $u_{min} = 90\mu\text{V}$ which is therefore the smallest modulator input voltage which can be sensed. Both these values were listed in Table 7.1.

2.2 LATCH

Two different modulators were designed: the “NRZ modulator”, which has NRZ and half-delayed NRZ (HNRZ) feedbacks, and the “RZ modulator” with RZ and HRZ feedbacks. Block diagrams of the latching schemes in each are shown in Figures 7.15a and 7.15b. The NRZ and RZ blocks are half-latches with outputs that either don’t or do return to zero after a half cycle (c.f. the

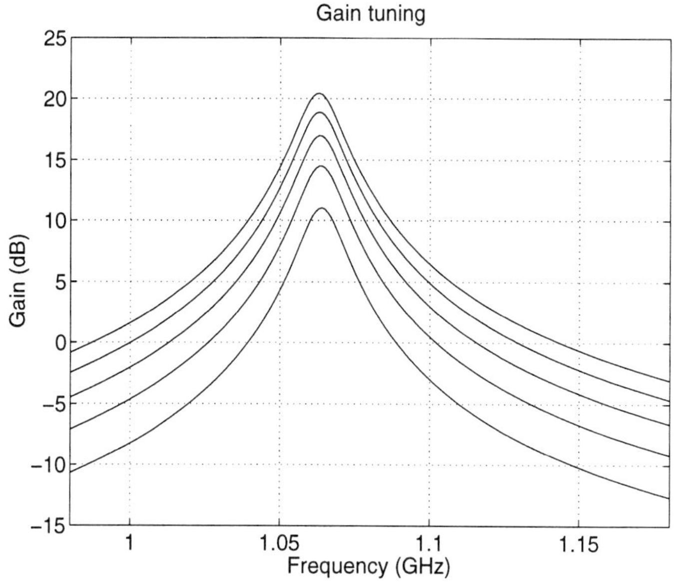

Figure 7.13a. Filter gain characteristic: fixed $V_Q = 2.6$V, stepping V_G by 0.2V from 2.0V to 2.8V.

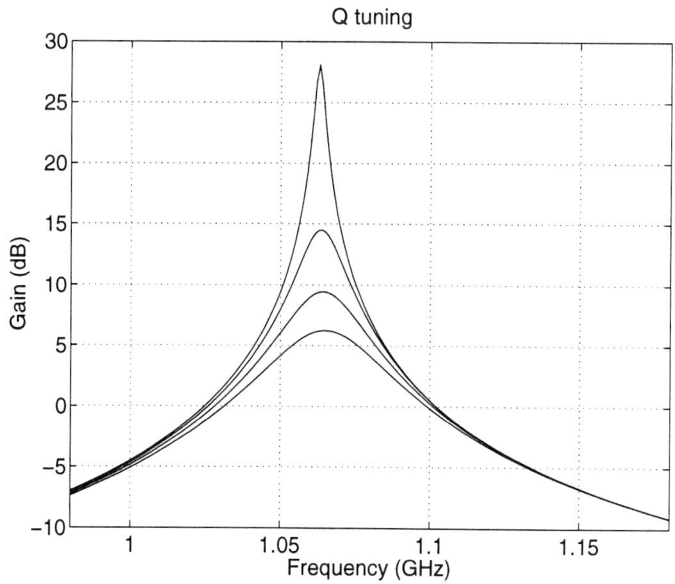

Figure 7.13b. Filter gain characteristic: fixed $V_G = 2.2$V, stepping V_Q by 0.1V from 2.4V to 2.7V.

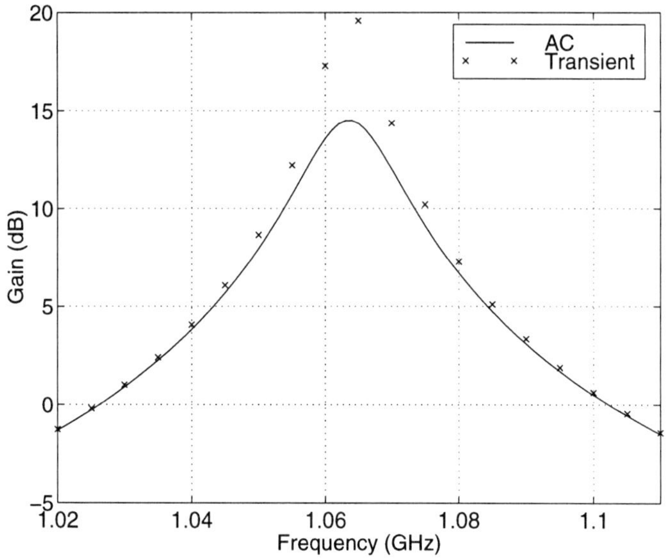

Figure 7.14a. Transient gain analysis for $V_G = 2.2\text{V}$ and $V_Q = 2.6\text{V}$.

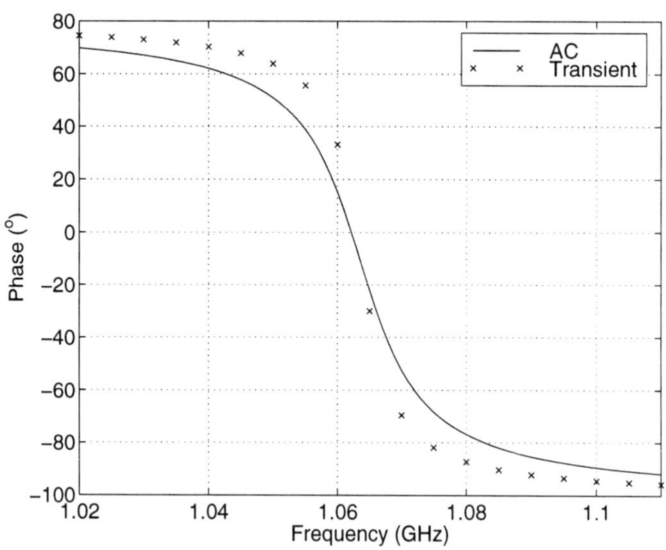

Figure 7.14b. Transient phase analysis for $V_G = 2.2\text{V}$ and $V_Q = 2.6\text{V}$.

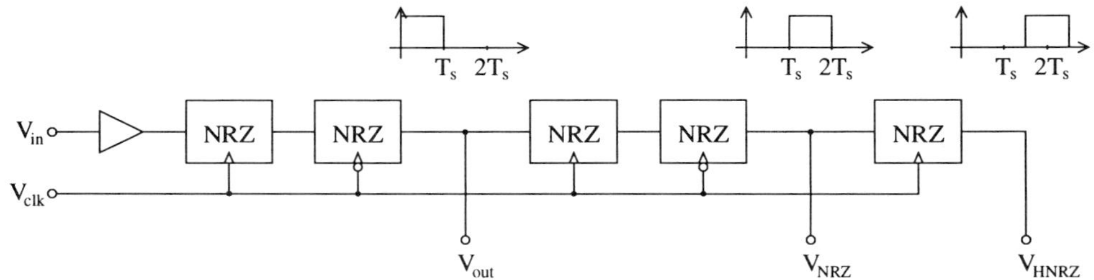

Figure 7.15a. Overall latching scheme and waveforms for NRZ modulator.

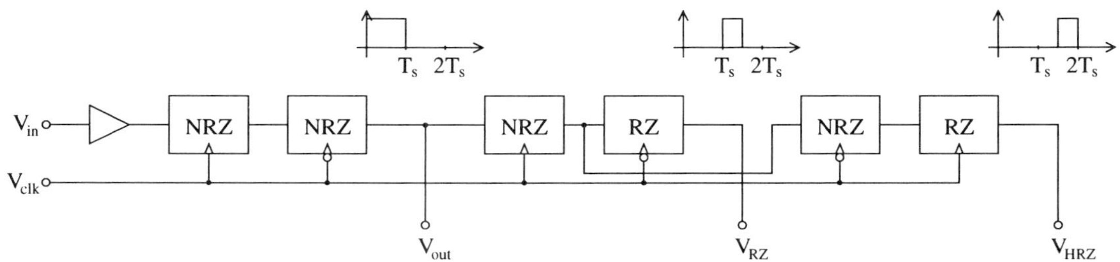

Figure 7.15b. Overall latching scheme and waveforms for RZ modulator.

dashed and dotted lines in Figure 2.6). The one-bit quantizer in both designs is a preamplifier with an ECL master/slave latch, exactly as was depicted in Figures 2.6 and 5.10. As mentioned at the start of this chapter, there is a full sample of delay prior to each DAC, and these are implemented with appropriate further half stages as shown.

The preamplifier circuit, Figure 7.16a, is a simple differential pair with input and output buffers. As we noted in Chapter 6 §6.3, it has been shown [Lee92] that emitter followers before and after a preamplifier are good for eliminating coupling of the clock signal between the main latch and the preamp. But the followers in this circuit were apparently sized and biased incorrectly for optimum speed, a rather serious oversight for a high-speed circuit. Gain and phase curves from SPICE ac analysis (supplemented with transient analysis for verification) for the input follower are shown in Figure 7.17a. Followers are fairly forgiving circuits, so the gain and phase shifts are not huge, but they can be improved a good deal. One result of the rolloff exhibited in the input follower gain curve is that the overall preamp has an ac analysis given in Figure 7.17b: the -3dB frequency is quite low at 4.59GHz. The dc gain is 18.7dB, and the phase shift at $f_{-3\text{dB}}$ is $-61.6°$.

A half-latch in this design appears in Figure 7.16b. The output swing is typically 270mV and the regeneration time constant for the latch as designed is $\tau_{rg} = 17.4$ps, which again can be improved with proper follower design.

Typical ZCT characteristics for the M/S latch output and one-sample delayed latch output are illustrated in Figures 7.18a and 7.18b, respectively. Clocking at $f_s = 4$GHz in an $f_T = 50$GHz process is aggressive, to be sure, if the guideline of 4–5% in Chapter 6 §7. is to be believed. That guideline was for a three half-latch design, and this one contains four, so the regeneration at the final latch output is adequate, as is clear by the sharpness of the corners in Figure 7.18b. However, it is the hysteresis in combination with small quantizer input swing which will turn out to be a major problem in this design: we barely use 5% of the full-scale input range. Our redesign will address this.

2.3 OUTPUT BUFFER

The output is obtained from the slave stage of the M/S latch driving a differential pair with 50Ω load resistors, Figure 7.19, for matching to an off-chip 50Ω measuring device. The bias current, and hence the output swing, is controlled with a voltage V_{BUF} applied directly to the base of a current-source transistor. A typical desirable swing for us is 200mV, which requires a current of 4mA; this can be achieved by using $V_{BUF} = 3$V.

In Figure 7.20a, we show a typical output waveform from the M/S latch, and Figure 7.20b depicts the output buffer waveform. The latch output seems not to rise very sharply, but the edge of the output buffer waveform is much better-defined.

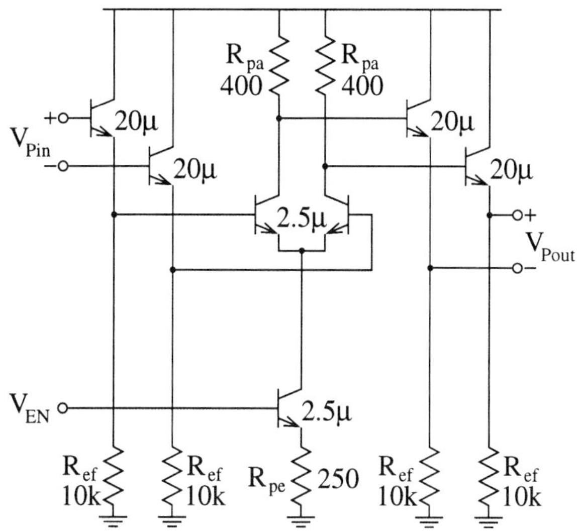

Figure 7.16a. Schematic for preamplifier.

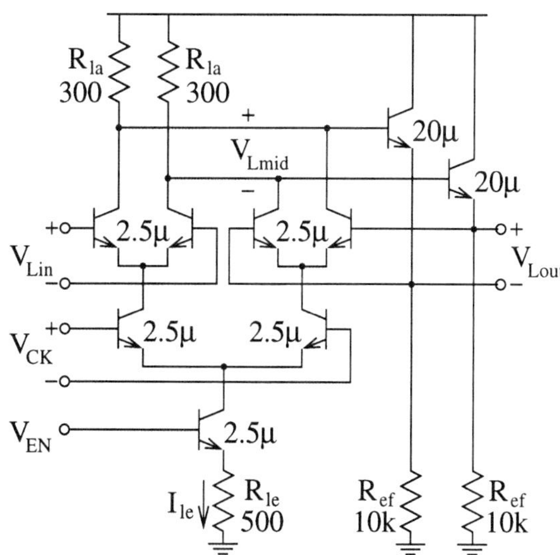

Figure 7.16b. Schematic for half-latch.

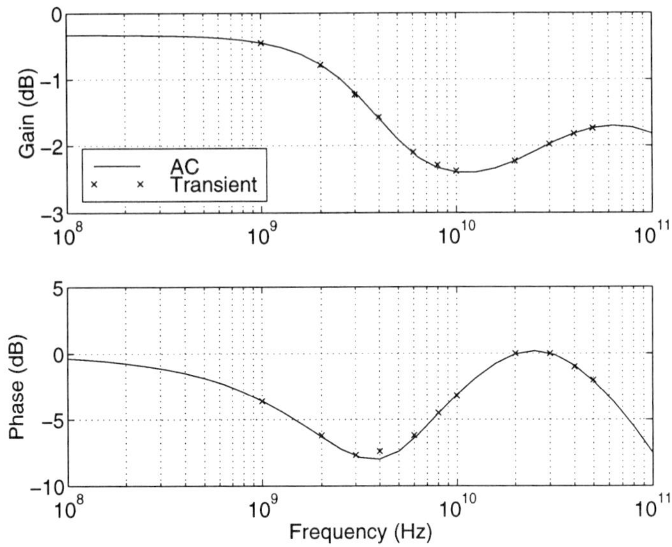

Figure 7.17a. Preamp ac analysis for input follower.

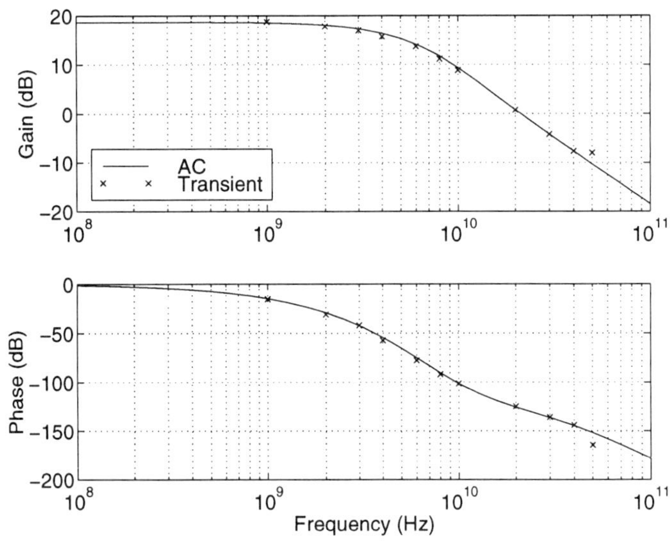

Figure 7.17b. Preamp ac analysis for entire circuit.

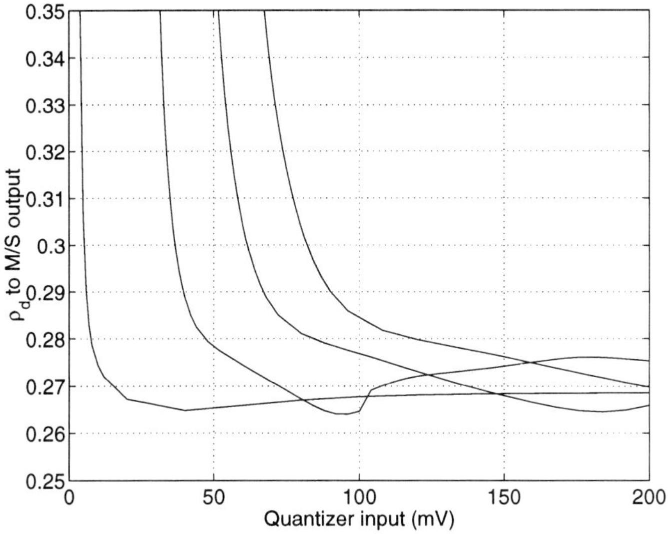

Figure 7.18a. ZCT characteristics for M/S latch output.

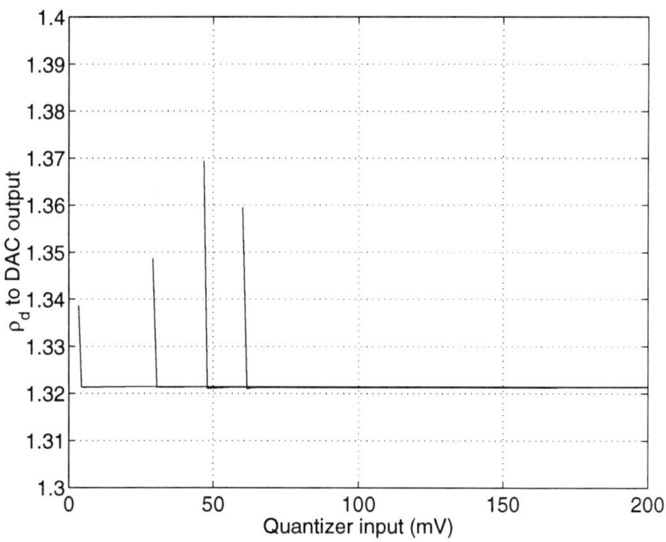

Figure 7.18b. ZCT characteristics for one-sample delayed output.

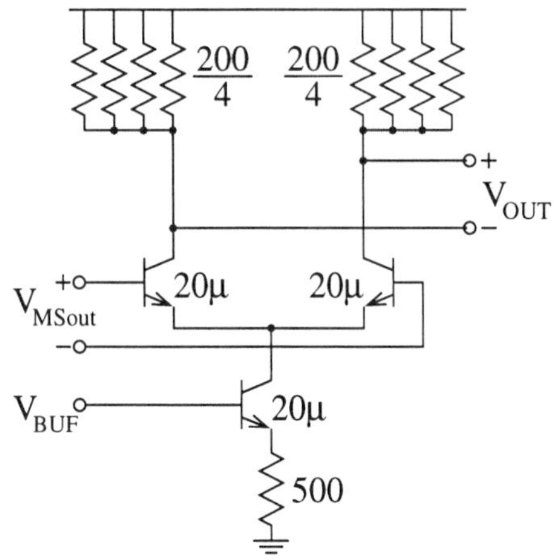

Figure 7.19. Modulator output buffer.

2.4 DAC

There are a total of four DACs on the chip, two for each of the V_{NRZ} and V_{HNRZ} (or V_{RZ} and V_{HRZ}) signals; for each signal, one DAC connects to the first resonator output and one to the second. The DACs are relatively simple current-steering circuits, Figure 7.21, with follower inputs and current outputs derived from the sum of cross-coupled differential pairs. A typical output voltage waveform vs. normalized time appears in Figure 7.22. The M/S latch output switches just after $t/T_s = 0$, then there is a full sample of delay before the DAC switches. The ZCT is $\rho_d = 1.322$ and the rise time is $\rho_r = 0.146$. Ideally, this waveform would switch instantaneously at $t/T_s = 1$.

2.5 COMPLETE CIRCUIT

Finally, complete transistor-level schematics for both modulators are shown in Figures 7.23a and 7.23b. The boldface words are the names of the external signals. A die photomicrograph of the fabricated NRZ modulator appears in Figure 7.24. The die measures $2.4 \times 1.6\text{mm}^2$ with the pads and $1.6 \times 0.85\text{mm}^2$ without. There are a total of 40 pads: 6 input (two each for differential input, clock, and output), 2 V_{CC}, 15 dc bias (V_{CB}, V_{BUF}, two each of V_G and V_Q, V_{CDB}, and four each of V_{k+} and V_{k-} for the DACs), and 17 ground. As is evident from the schematics, the input signal common-mode (CM) levels are not set on-chip and so must be provided through bias tees.

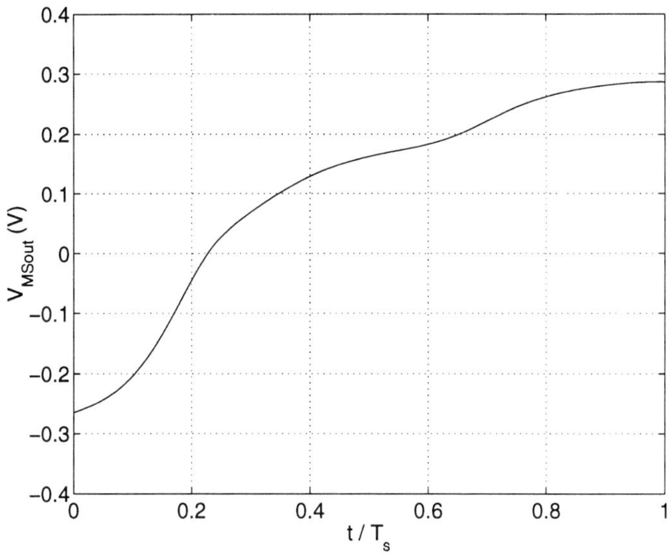

Figure 7.20a. Output waveform from M/S latch.

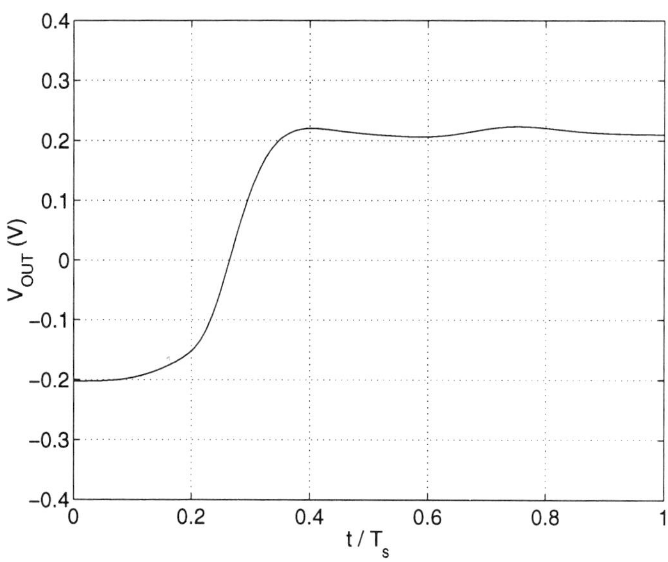

Figure 7.20b. Output waveform from output buffer.

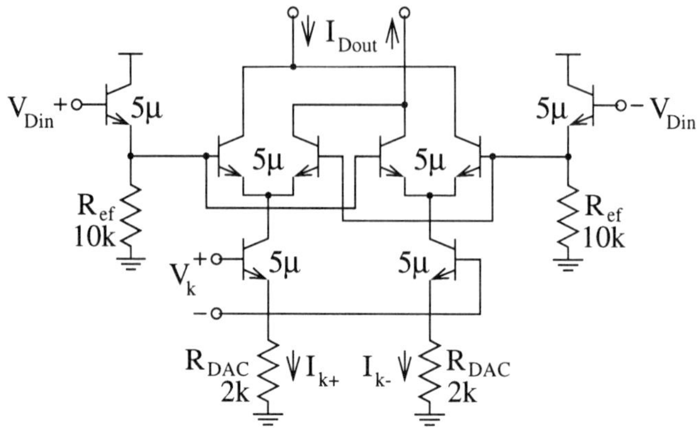

Figure 7.21. Current-steering DAC schematic.

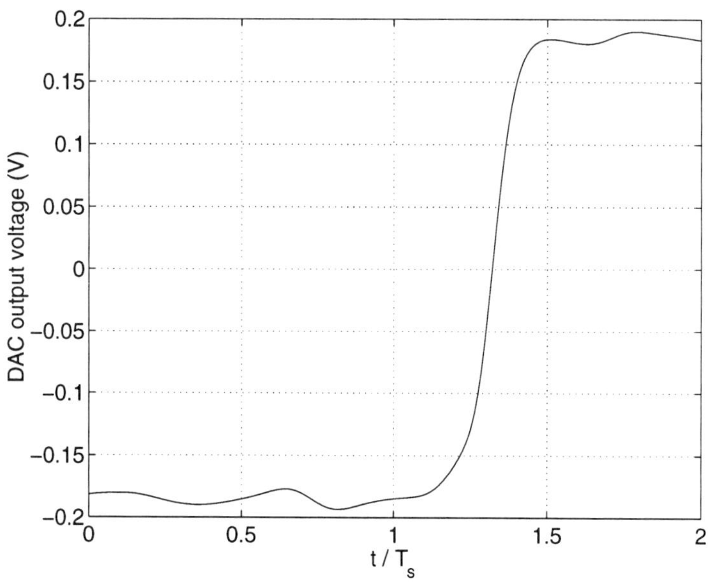

Figure 7.22. Dynamic DAC output voltage.

3. MEASUREMENT RESULTS

Standard high-speed probe configurations for wafer-level tests did not exist for this pad ring. It would have been possible to have a "membrane probe card" specially constructed, but several factors (not least the financial expense) ruled out this possibility. Thus, diced wafers were packaged and mounted on a four-

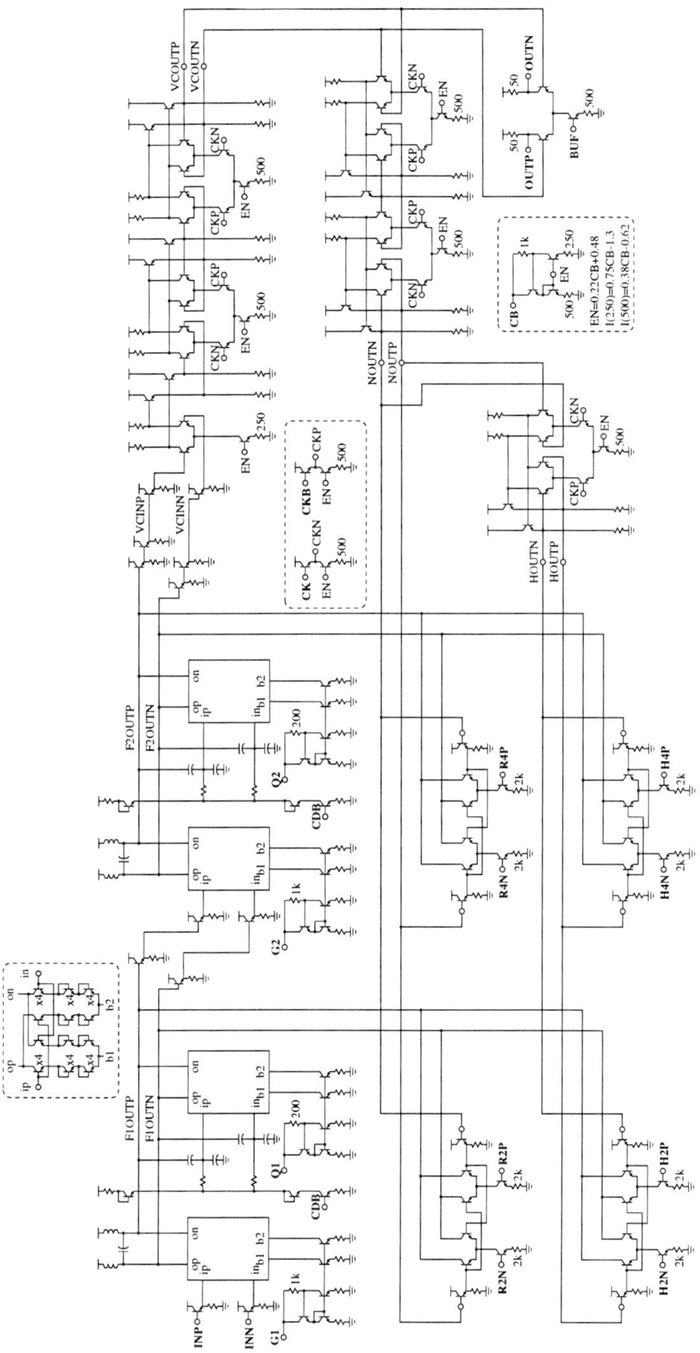

Figure 7.23a. Complete NRZ $\Delta\Sigma$M schematic.

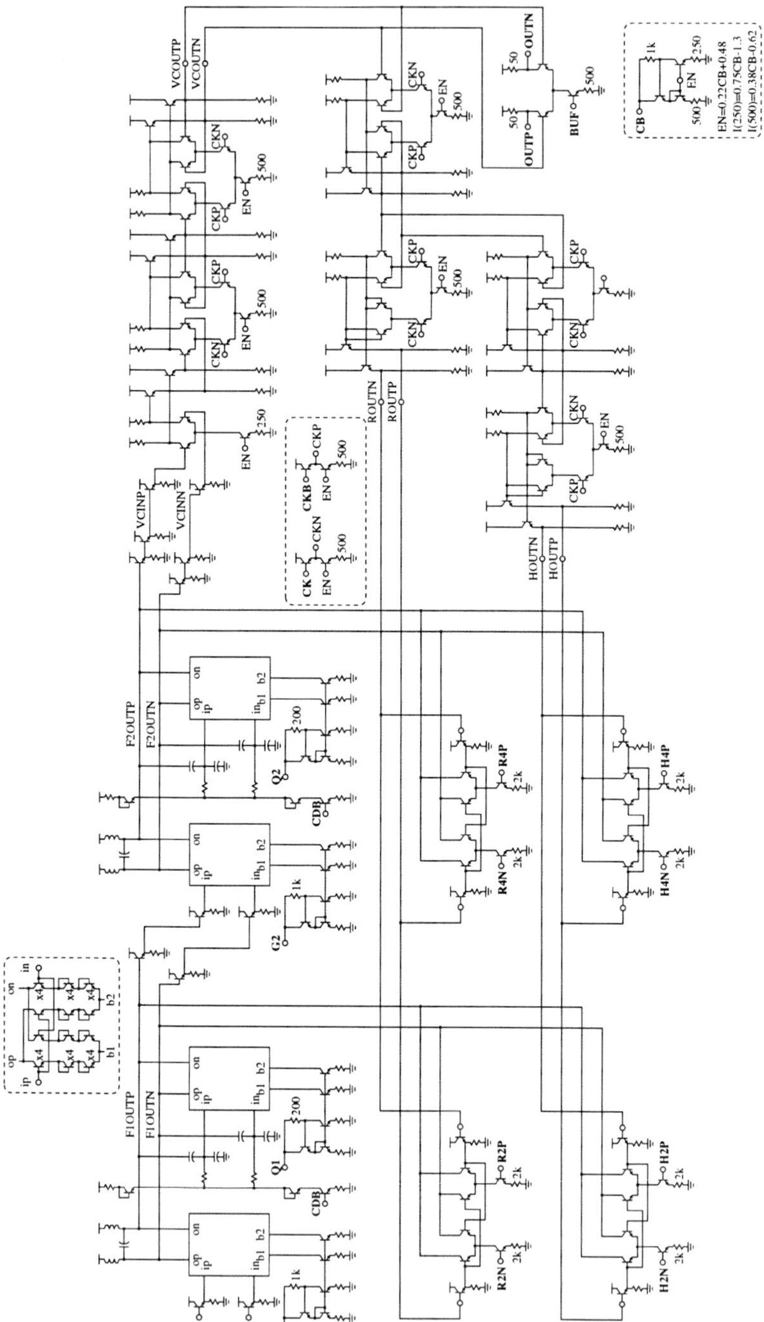

Figure 7.23b. Complete RZ ΔΣM schematic.

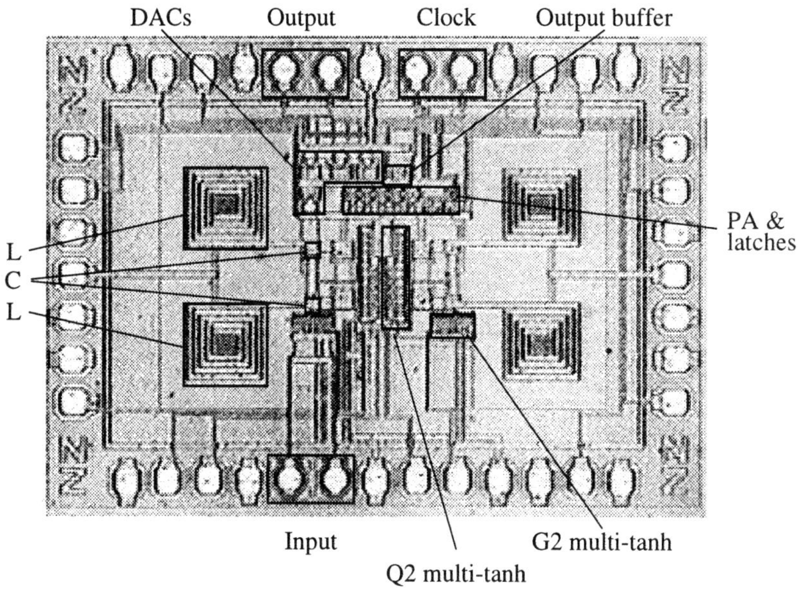

Figure 7.24. Die photomicrograph of NRZ modulator.

layer test board. There are no individual circuit block breakouts, so we must devise methods to check circuit behavior based only on the overall modulator output bit stream, either in the time domain or the frequency domain.

The input and clock signals were provided from signal generators, driven differentially onto the board through 180° power splitters, each with two bias tees for providing CM voltages. At first, V_{CC} and the two CM voltages were all provided with separate power supplies, which resulted in the destruction of several packaged parts by applying inappropriate voltages across certain junctions of the input buffer transistors; eventually, a single supply for V_{CC} was used, with the CM levels being drawn from a tunable resistive divider circuit. SMA connector-terminated cables of equal lengths rated to 40GHz were connected to the board. The remaining dc biases were set with 10kΩ potentiometers connected between V_{CC} and ground. Each required hand tuning with a screwdriver and voltmeter to set a desired voltage level. The output came differentially from two similar cables connected to the "DC+RF" input of bias tees; the RF outputs have no remaining dc component, and they were connected to a spectrum analyzer through a combiner. A diagram of the test setup appears in Figure 7.25.

A typical spectrum analyzer display for the NRZ modulator output bit stream appears in Figure 7.26. The conditions with which this plot was made were:

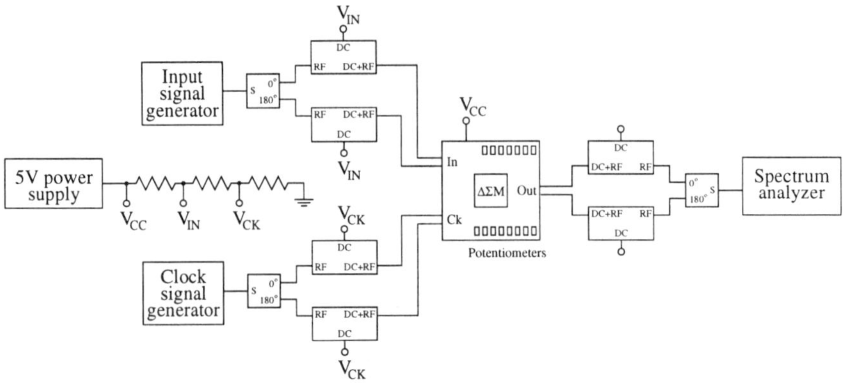

Figure 7.25. Measurement test setup.

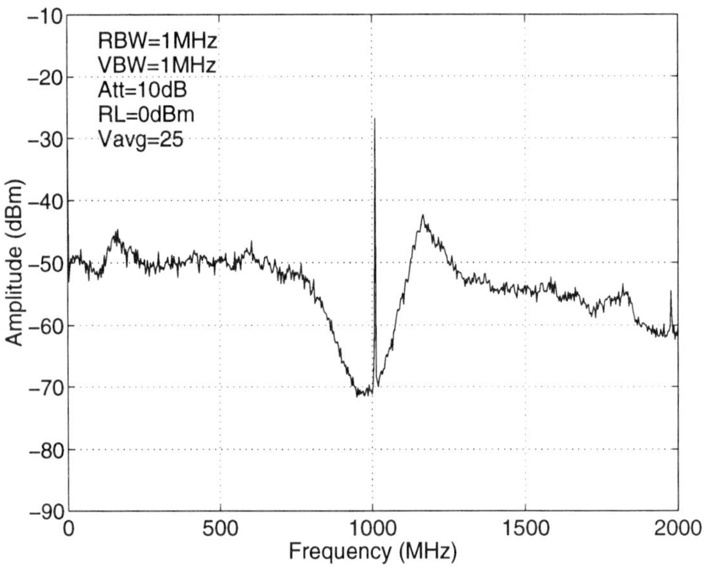

Figure 7.26. NRZ modulator output spectrum.

- Input 4.2V CM, −26dBm, 1.010GHz
- Clock 3.0V CM, −10dBm, 4.000GHz
- Power supply 5.03V, 75mA
- $V_{G1} = V_{G2} = 2.6\text{V}$, $V_{Q1} = V_{Q2} = 2.8\text{V}$, $V_{CDB} = 0\text{V}$

- $V_{BUF} = 3\text{V}$, $V_{CB} = 4\text{V}$
- DAC CM 1.2V, $V_{n2} = 0.4V$, $V_{n4} = 0.2\text{V}$, $V_{h2} = V_{h4} = 0\text{V}$

The noise-shaping behavior is evident: the quantization noise has a dip of 20dB or so at about $f_s/4 = 1\text{GHz}$. Thus, the circuit appears to be functioning correctly.

3.1 RESONATOR

For the resonator, there are three voltages which should have a noticeable effect: V_G, V_Q, and V_{CDB}. Let us examine each in turn.

VARYING V_G

For simplicity, we test the NRZ modulator with the half-delayed feedback pulses disabled, i.e., $V_{h2} = V_{h4} = 0\text{V}$. This makes it easy to specify the level of a full-scale input. From (6.14) and Figure 6.1 (recall also Example 1.7), the maximum input is one where the current due to the input transconductor $G_{g1}u$ has the same magnitude as the feedback current $k_{n2}y$. Thus, the output magnitude relative to full scale of an input signal u in V is

$$\frac{G_{g1}}{k_{n2}}u, \tag{7.4}$$

where G_{g1} is in mA/V and k_{n2} in mA. Let us relate all these quantities to the signals we actually use.

- G_{g1} is related to V_{G1} through Figure 7.A.2a. An approximate formula is

  $$G_g = 5.43V_G - 8.73, \tag{7.5}$$

  where G_g is in mA/V and V_G in V.

- For k_{n2}, define $k_{n2} \equiv I_{k+} - I_{k-}$ and $V_{n2} \equiv V_{k+} - V_{k-}$ in Figure 7.21. From Figure 7.A.4,

  $$I_k = 0.48V_k - 0.38 \tag{7.6}$$

  for I_k in mA and V_k in V. Substituting our definitions in (7.6), we arrive at

  $$k_{n2} = 0.48V_{n2} \tag{7.7}$$

  for k_{n2} in mA and V_{n2} in V.

- The input voltage V_{in} from the signal generator is calibrated in dBm assuming a 50Ω load. The modulator input, however, was not designed to have a 50Ω input impedance: the signal generator drives the pin capacitance, a bond wire inductor, and then an emitter follower. Using SPICE, we estimate that the signal at the base of the emitter follower will be about 4dB smaller than the dBm reading on the generator. Taking this into account, we may write

$$u = 10^{(V_{in}-14)/20} \tag{7.8}$$

where u is in peak V (as opposed to rms) and V_{in} is the nominal output from the generator in dBm.

- We shall see that the time-domain output voltage has a swing of about 120mV peak-to-peak, or 60mV peak. An 0dB input tone thus requires a peak magnitude of 60mV, which is

$$20 \log_{10} 0.06 + 10 = -14.5\text{dBm}. \tag{7.9}$$

For a 1.003GHz -26dBm input, the magnitude of the 1.003GHz tone in the output spectrum as a function of V_{G1} appears as the solid line in Figure 7.27a. The result calculated from equations (7.4)–(7.9) is plotted as the dashed line. The curves agree reasonably well. According to our approximate formula (7.5), the transconductor turns off at $V_{G1} = 1.6$V, at which point no output tone should be visible; in reality, the output tone remains with an amplitude of about -47dBm even with $V_{G1} = 0$V, likely because of coupling across C_μ from input to output of G_{g1}.

Simulation of the modulator using the RK4 program and a model like Figure 6.1 shows that the output amplitude depends slightly on V_{G2} as well, though in a manner that is more difficult to calculate. Figure 7.27b shows measured and simulated output tone magnitude against V_{G2} for fixed $V_{G1} = 2.6$V, and once again, even with $V_{G2} = 0$V, there is still an output tone of about -35dBm. Generally, the output tone magnitude behaves as expected when either V_G is varied.

VARYING V_Q

Ideally, a modulator is tuned so that it has infinite Q. If it were, it would have an infinitely-deep notch in the quantization noise and optimal SNR. Practically, there are two cases of interest which we demonstrate here. If Q is tuned too low, then the modulator will not exhibit noise shaping for very small input levels [Fee91]. Figure 7.28a depicts output spectra for $V_{Q1} = V_{Q2} = V_Q = 2.8$V and the input amplitude increasing slowly. At -48dBm, no noise shaping is seen; the modulator is sensitive only to inputs of -47dBm. Figure 7.28b plots the minimum-detectable input amplitude versus V_Q: at $V_Q = 3.25$V, the input can be disabled without the loss of noise shaping at the output.

On the other hand, if Q is too high, then instead of a resonator we will have an oscillator. Figure 7.29a is the output spectrum for $V_Q = 3.31$V

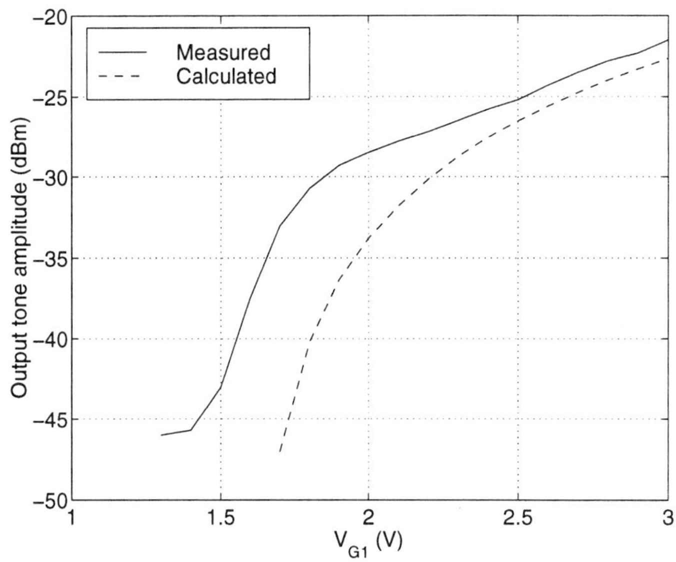

Figure 7.27a. Tone magnitude in output spectrum against V_{G1}.

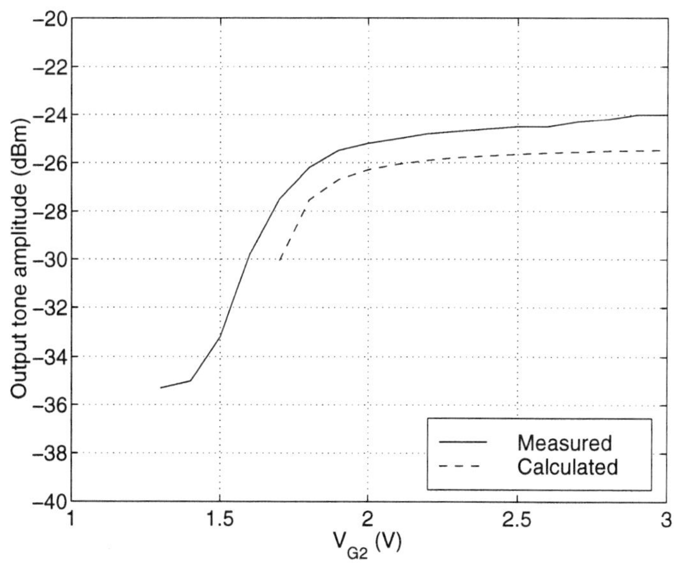

Figure 7.27b. Tone magnitude in output spectrum against V_{G2}.

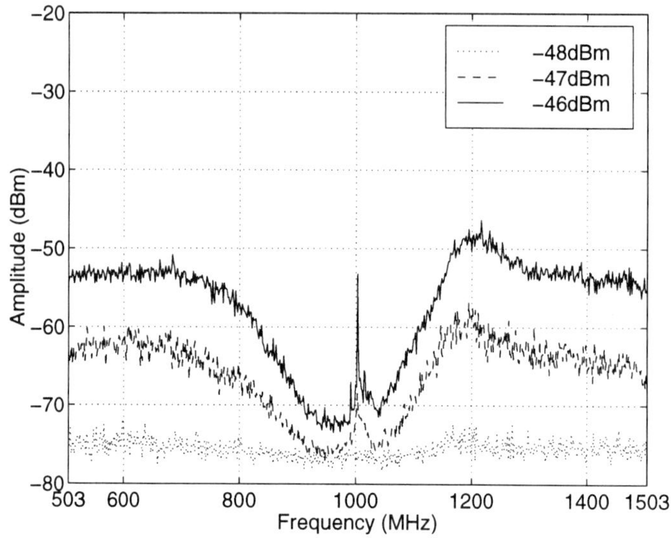

Figure 7.28a. Loss of noise shaping for small input and low Q.

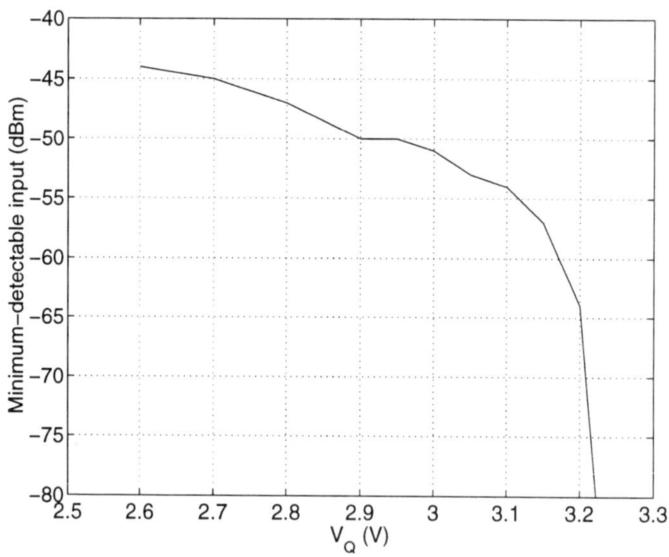

Figure 7.28b. Minimum-detectable input amplitude against V_Q.

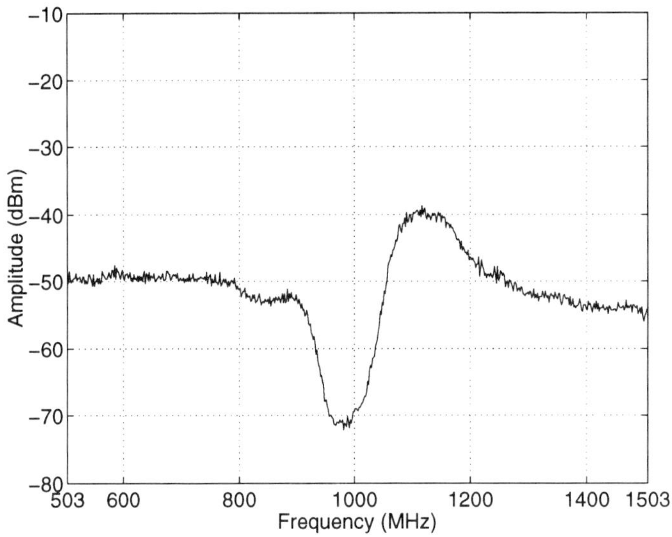

Figure 7.29a. Output spectrum just before oscillation ($V_Q = 3.31$V).

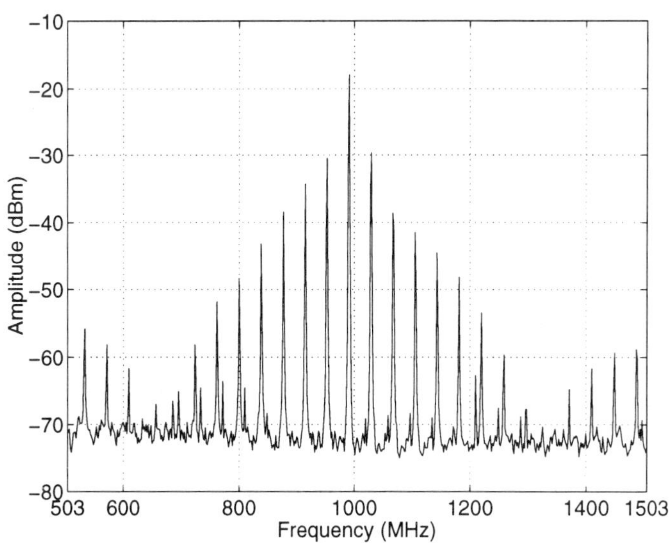

Figure 7.29b. Output spectrum just after oscillation ($V_Q = 3.33$V).

and no input, which turns out to be just on the edge of stability—if we make $V_Q = 3.33$V, the spectrum looks like Figure 7.29b, which has the tonal behavior characteristic of a modulator with an oscillator in the forward path[2]. The V_Q which causes oscillation in practice is higher than that predicted by (6.14): from Figure 7.A.2b, we may write

$$G_q = 5.61V_Q - 9.07 \tag{7.10}$$

for G_q in mA/V and V_Q in V, and for the parameters in Table 7.1, we expect $V_Q = 2.38$V or so. The spacing of the tones in Figure 7.29b is curious: they seem to occur every 38MHz or so, which suggests we have entered an output limit cycle whose period is $4\text{GHz}/38\text{MHz} \approx 105$. No explanation for this value is obvious, though it might reasonably be some kind of beating between the oscillation frequency and $f_s/4$. Naturally, the modulator is not intended to be operated in this regime.

VARYING V_{CDB}

We stated in §2.1 that setting V_{CDB} too high would give linearity problems due to saturating the current-source transistors in the Q-enhancement multi-tanh block. The best way to verify this in practice would be with a two-tone test, but our setup is too cumbersome to allow this to be done easily. However, with a fixed-amplitude −26dBm tone, we can observe the amplitude of the output tone varying as V_{CDB} increases, Figure 7.30a. The gain to the output is constant for small V_{CDB}, but as soon as Q_{CDB} in Figure 7.12 turns on, we start to see distortion, first gain expansion, then gain compression. The transistors supplying I_{tail} in Figure 7.7 are being driven into saturation almost immediately when Q_{CDB} starts to conduct: Figure 7.30b shows that the current drawn from the supply begins to drop as V_{CE} is driven towards 0, which is the expected behavior from Figure 7.A.1b.

3.2 LATCH

The control voltage V_{CB} affects the behavior of the latch in a quantifiable way: it changes the current I_{le} through R_{le} in Figure 7.16b, and hence the regeneration time τ_{rg}. Increasing I_{le} leads to a closer bunching of the ZCT curves in Figure 7.18a, in turn leading to smaller σ_{DPW}. Examination of the output spectrum near $f_s/4$ shows that the spectrum is white; if it is limited by noise due to DPW modulation from metastability, then the noise floor should

[2] It is actually possible to have a pole slightly outside the unit circle and still have a stable modulator: this yields a so-called chaotic modulator [Ris94, Chap. 3]. Pushing the pole too far outside the unit circle results in instability like that shown in Figure 7.29b. From Figure 7.28b, $Q = \infty$ seems to be achieved at $V_Q = 3.22$V, so the plot in Figure 7.29a for $V_Q = 3.31$V is very likely one where the modulator is chaotic.

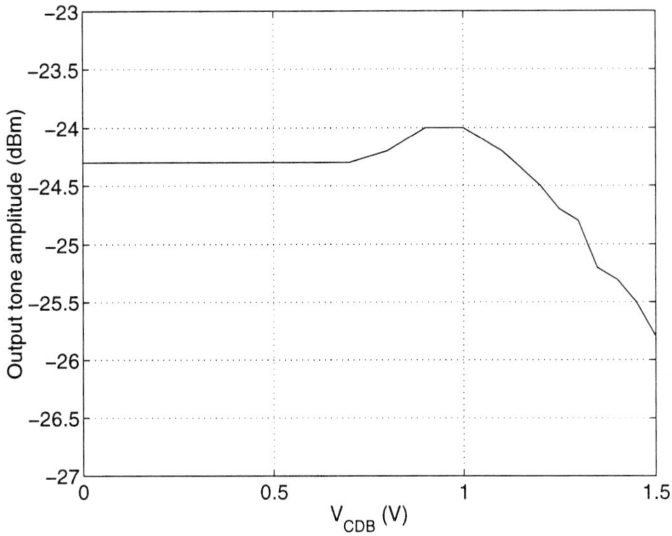

Figure 7.30a. Effect of V_{CDB} on linearity.

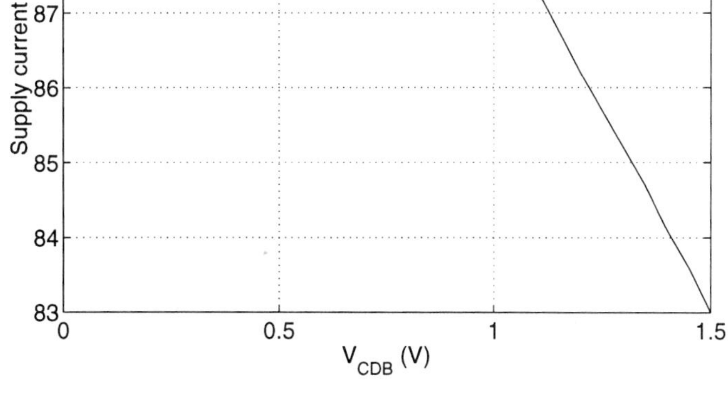

Figure 7.30b. Effect of V_{CDB} on supply current.

become lower as σ_{DPW} falls. Figure 7.31a demonstrates that the noise floor near the resonator center frequency does indeed fall as V_{CB} is increased from 2.6V to 3.3V and 4.0V. Figure 7.31b is the in-band noise measured with the spectrum analyzer set to display a 10MHz band near the approximate notch center frequency of 980MHz; as V_{CB} increases, the in-band noise falls.

We had the opportunity to take some time-domain measurements on a 50Gs/s sampling oscilloscope. Figure 7.32 shows an eye diagram of the 4Gb/s bit stream on a 50ps/div horizontal time scale. The eyes have a not-inconsiderable number of dots inside them, the cause being the location at which the output bits are taken: the M/S latch output. The dots correspond to instances where the latch output is delayed and the oscilloscope happens to sample at a point on that delayed waveform. The eyes would be more open if the output bits were taken from the one-sample delayed latch, where from Figure 7.18b there is less ZCT variance. This is of consequence when the modulator output spectrum is examined on a spectrum analyzer, as opposed to being found from the FFT of a sequence of output bits. Even though much of the digital output edge jitter due to quantizer metastability is removed in the *feedback path* by the two extra half-latches, none of this jitter is removed at the modulator output because this output is taken *prior* to the extra regeneration stages. To a spectrum analyzer, the analog properties of the output waveform are significant; thus, closed eyes due to metastability will degrade the spectrum measured on a spectrum analyzer by whitening it in-band. We *did* find that increasing V_{CB} resulted in improved eye openings, as one would expect.

More evidence of the effects of metastability can be seen in Figure 7.33, which is a histogram of the zero crossings of the time-domain output bit stream. At these rapid speeds, the sampling jitter of the oscilloscope itself is significant. When we used a common frequency reference for both a signal generator and the scope, and applied a 4GHz sine wave from the generator to the scope, we found a normally-distributed time jitter (the "RMSΔ" field in the figure) in the sine wave zero crossings of about 7ps. For the $\Delta\Sigma$M output, the rms jitter is 18.7ps, and there it is clear that the tail of the distribution descends more gradually to the right than the left as expected for a metastable quantizer. The same measurement was taken five times, and rms jitters of $\{18.2, 18.3, 18.7, 19.2, 21.2\}$ps were measured, so it is hard to specify an exact value.

In theory, the effect of metastability could be removed if we captured a bit stream and found its spectrum. We briefly had access to an 8Gs/s oscilloscope which could sample and hold 128k data points. The scope thus sampled the 4Gs/s bit stream twice per bit, and then the odd or even 64k samples could be downloaded to a computer and a spectrum could be taken. For a 1GHz -15dBm input, Figure 7.34 shows the 16 averaged 4096-point Hann-windowed periodograms of the bit stream. Annoyingly, this data was captured with biases

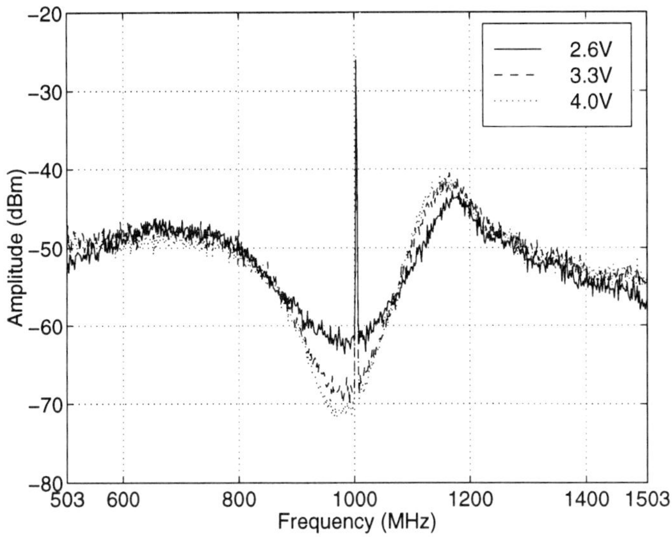

Figure 7.31a. Effect of V_{CB} on overall spectrum.

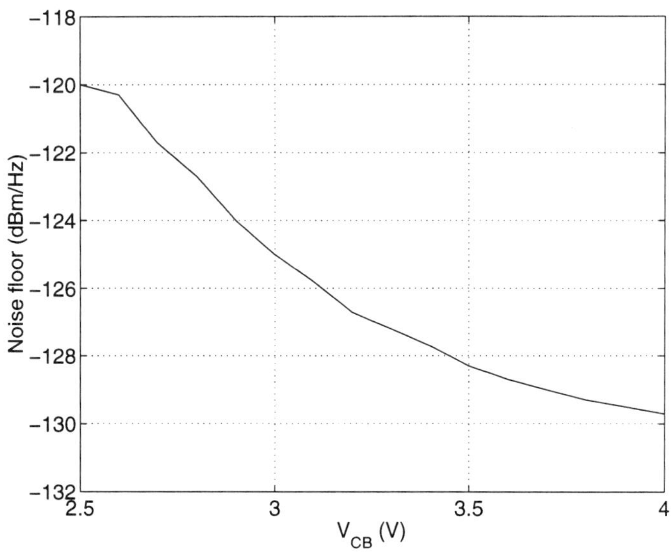

Figure 7.31b. Effect of V_{CB} on in-band noise.

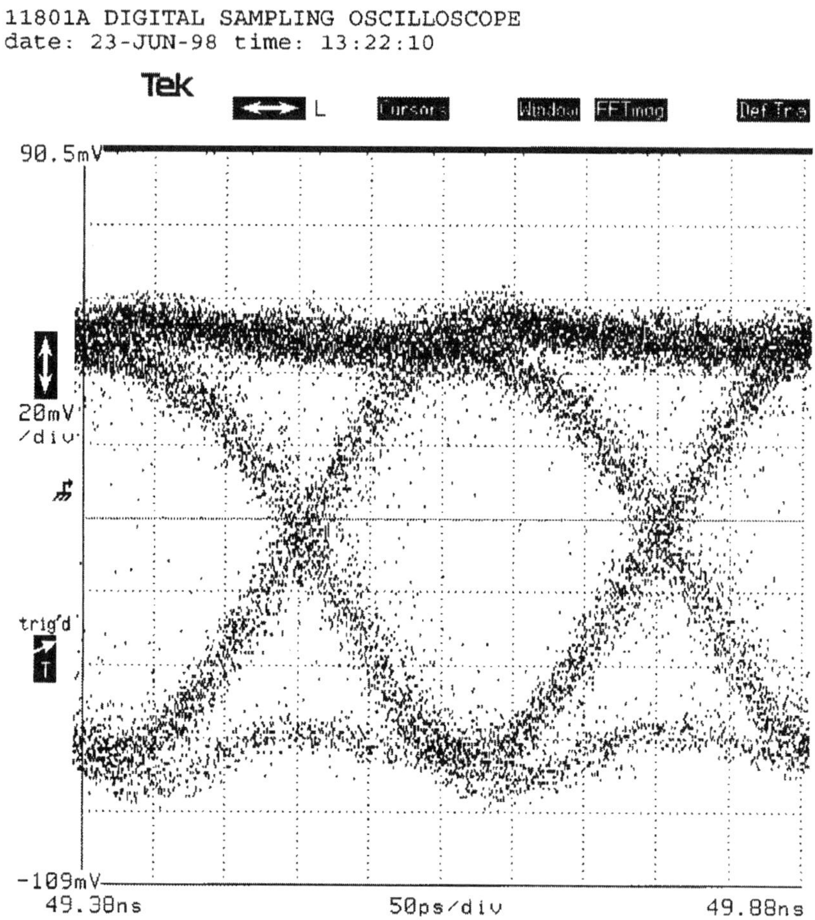

Figure 7.32. Output bit stream eye diagram.

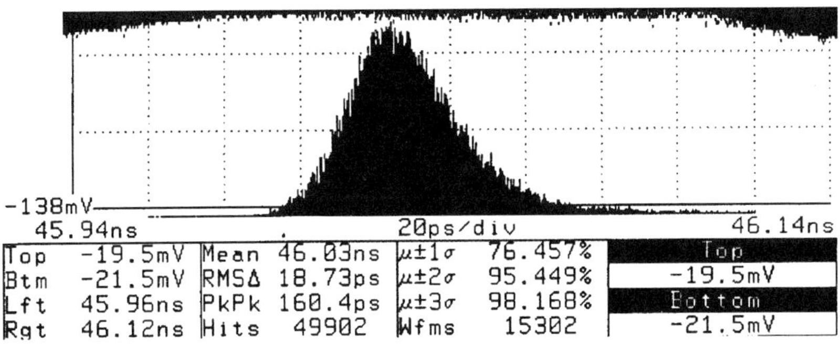

Figure 7.33. Histogram of time-domain output bit zero crossings.

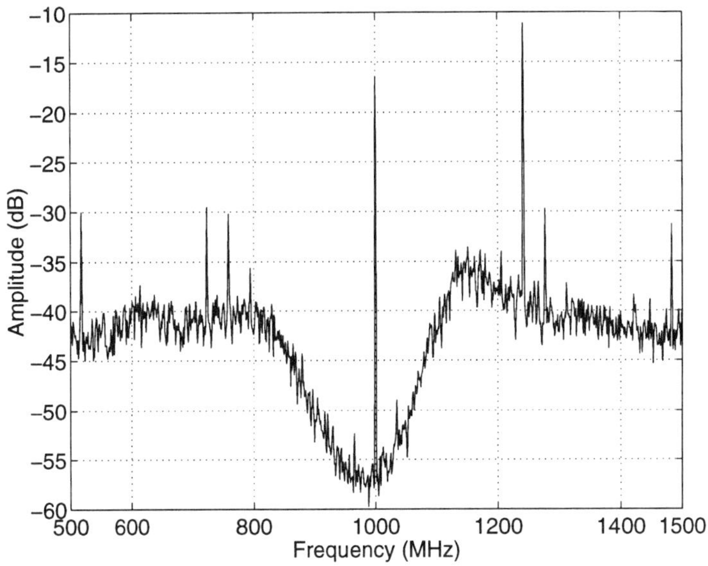

Figure 7.34. Output spectrum of captured bit stream.

set incorrectly: we used $V_Q = 2.6$, which is nowhere near high enough for a deep noise notch. As a result, this spectrum has a similar noise floor to that observed on the spectrum analyzer and many spurious tones.

3.3 OUTPUT BUFFER

There is not much to test on the output buffer itself. Figure 7.35 plots output spectra for $V_{BUF} = 2.0$V and 3.0V. From Figure 7.19, such a change should result in a supply-current increase from about 2mA to about 4mA,

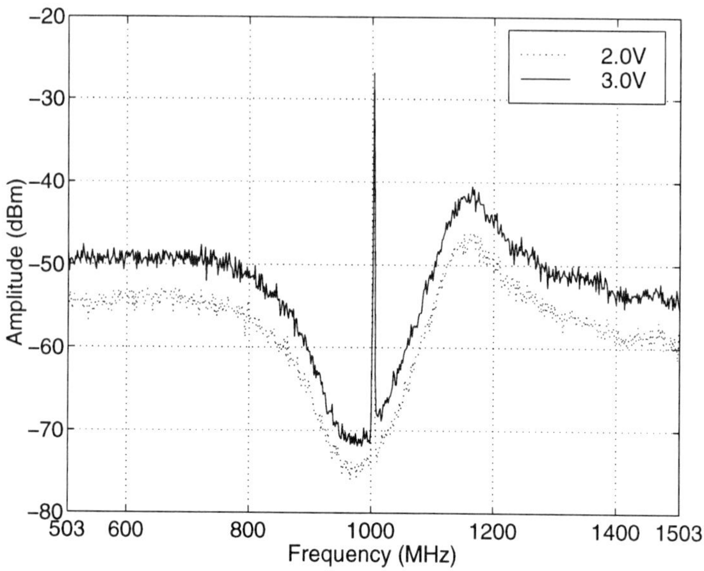

Figure 7.35. Effect of V_{BUF}.

and we measure it to go from 87.8mA to 90.0mA. Furthermore, doubling the switching current should mean double the output voltage swing and hence 6dBm more spectral power total; the measured increase is about 5dBm.

3.4 DAC

Changing V_{n2} in the NRZ modulator should affect the amplitude of the input tone as it appears in the output spectrum according to (7.4). Figure 7.36a is the output magnitude of a 1.003GHz -36dBm input tone as V_{n2} varies, both measured and calculated. Once again the shapes of the curves agree quite well. The supply current rises approximately 400μA over the range $V_{n2} = 0.2$V to 1.0V, which concurs with Figure 7.A.4.

Otherwise, changing DAC voltages should affect the loop filter zero locations and hence the noise shaping behavior. Figure 7.36b shows that an NTF pole moves up in frequency and further towards the middle of the z plane as V_{n2} is increased; a similar movement of this pole is observed in simulation in the RK4 program. Generally, pole movement is as expected.

3.5 DYNAMIC RANGE

It is clear from the spectra presented in this section that the actual noise notch center frequency is approximately 980MHz. Thus, for a DR plot, we

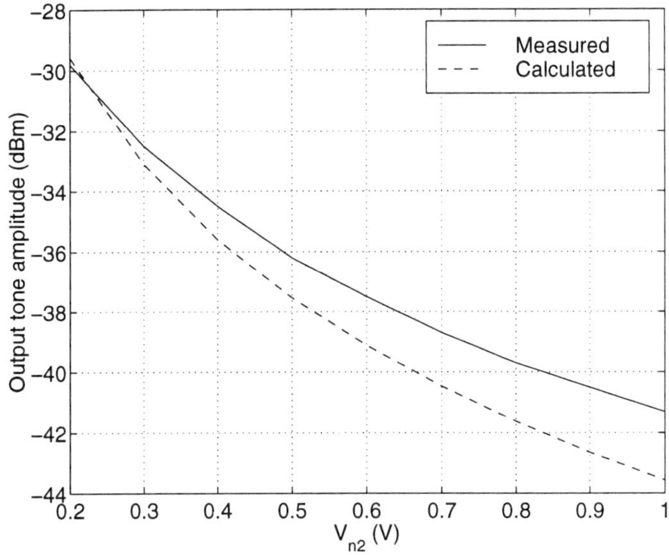

Figure 7.36a. Effect of changing V_{n2} on tone magnitude at output.

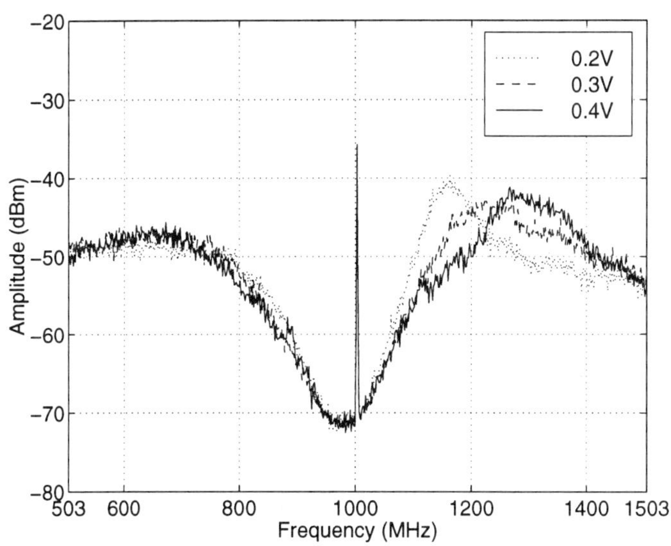

Figure 7.36b. Effect of changing V_{n2} on overall spectrum.

set $f_s = 3920$MHz, four times this value, and apply an input tone at 981MHz. We choose to consider a bandwidth of 20MHz (i.e., OSR $= 100$), which is what we assumed in Table 7.1. Typical in-band spectra appear in Figures 7.37a and 7.37b; the noise floor is white with a level of about -130dBm/Hz. In a 20MHz bandwidth, the total noise power is thus -57dBm, and in Figure 7.37a, the signal power is -42dBm, which gives SNR $=$ 15dB. An input 20dBm larger shows much output harmonic distortion, as seen in Figure 7.37b. From (7.4) and the circuit voltages, the full-scale input for these conditions is about -14dBm.

Figure 7.38 shows the DR plot for the NRZ modulator, where spectra with 20MHz bandwidth were captured and the SNR calculated in Matlab. This SNR stays positive for inputs up to 0dB, but it is difficult to know how to calculate it fairly for large inputs because of strong harmonic content and distortion of the input signal. In any event, the peak SNR is 37dB and the DR is about 40dB, which makes this a 6.3-bit converter. The SNR value agrees well with our prediction in Table 7.1. The modulator consumes about 450mW from a single 5V supply. With a -5dB input, the harmonic at 977MHz dominates, and the SFDR is found to be approximately 48dB. In a narrower bandwidth, SNR performance would be better, improving at 3dB per octave of oversampling, though spurs can exist even in very narrow bands and SNR might still be limited.

4. RESULT COMMENTARY

To some degree, we were hampered by the fact that during the design phase we did not anticipate how we would test this modulator. Circuit blocks were not broken out individually, and package and board design had to be done carefully.

Even without these problems, clock jitter appears to be the major limit to the observed performance. Some further comments about this are in order.

- One source of jitter might be the clock signal source. We measured its phase noise to be better than -95 dBc/Hz at 100 kHz offset from the 4 GHz carrier; according to our results in Chapter 5, this is unlikely to be high enough to limit the performance to the degree we observed.

- Even if the clock were to have no phase noise, the input-referred noise voltage at the base of the clock input buffer would cause a deviation in the zero-crossing of the clock voltage, i.e., clock jitter [vdP94]. In Figure 7.39, near a zero crossing of the clock waveform, circuit noise Δv_{noi} causes a time jitter Δt_{jitt}. For a clock given by

$$V_{clk} = A_{clk} \sin 2\pi f_s t, \tag{7.11}$$

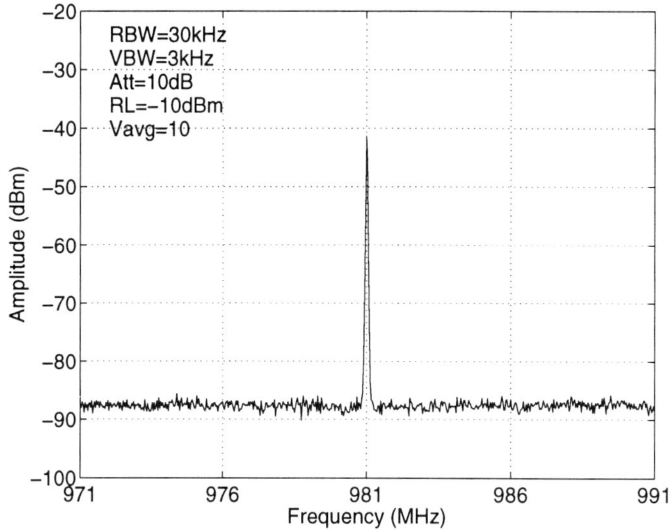

Figure 7.37a. Measured modulator spectrum for $V_{in} = -40$dBm.

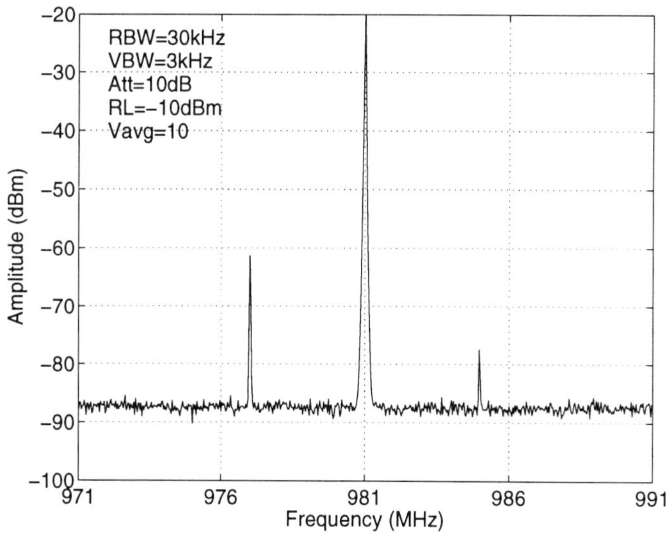

Figure 7.37b. Measured modulator spectrum for $V_{in} = -20$dBm.

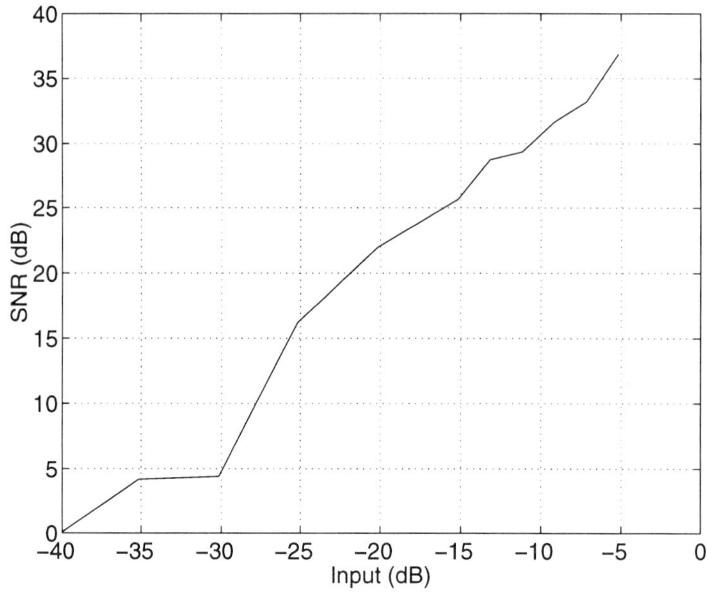

Figure 7.38. Dynamic range plot for NRZ modulator in 20MHz bandwidth.

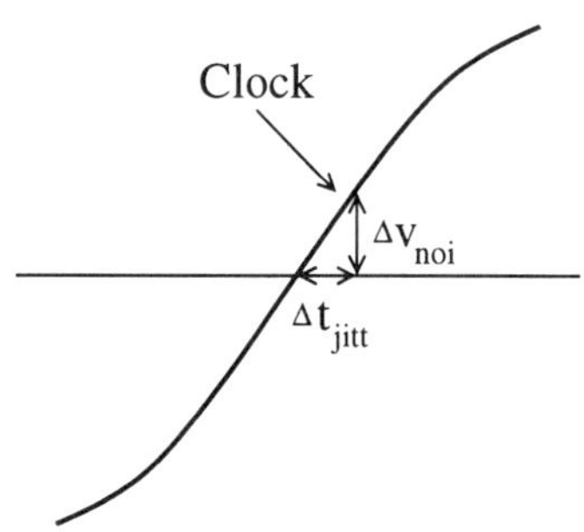

Figure 7.39. Clock jitter caused by circuit noise.

Figure 7.39 allows us to write

$$\frac{\Delta v_{noi}}{\Delta t_{jitt}} \approx \left.\frac{dV_{clk}}{dt}\right|_{\max} \tag{7.12}$$

because the clock zero crossings are at the point of maximum slope of the clock waveform. Solving for Δt_{jitt} gives

$$\Delta t_{jitt} = \frac{\Delta v_{noi}}{2\pi A_{clk} f_s}; \tag{7.13}$$

writing $\Delta v_{noi} \equiv \sqrt{v_{nc}^2 f_s}$ and solving yields

$$\Delta t_{jitt} = \frac{\sqrt{v_{nc}^2 f_s}}{2\pi A_{clk}} T_s. \tag{7.14}$$

In our modulator, SPICE ac analysis tells us that the input-referred noise voltage at $f_s = 4$ GHz for the clock transistor is $\overline{v_{nc}} = 3.56$ nV/$\sqrt{\text{Hz}}$, so for $A_{clk} = -4$ dBm $= 0.2$ V,

$$\Delta t_{jitt} = 1.8 \times 10^{-4} T_s = 45 \text{ fs}. \tag{7.15}$$

This is also too low to be responsible for the observed performance.

- Our clock input pads went directly to the bases of clock buffer transistors whose impedances were not matched to 50 Ω. Simulations indicate this lowers A_{clk} at the output of these buffers by about 10 dB; this would raise t_{jitt} in (7.15) by a factor of 3, which is still not enough to explain the observed performance.

- When a zero-crossing histogram like Figure 7.33 was plotted for just the 4 GHz clock, its 1σ value was 3.5 ps or so; this might be attributable to the 10 MHz reference clock used as a trigger on the high-speed sampling scope.

Unfortunately, we are unsure of the main reason for the observed jitter.

In fairness, it must be noted that a spectrum analyzer is not the best way to characterize the performance of a high-speed $\Delta\Sigma$M because, as we explained, it is sensitive to analog imperfections in the output bit stream waveform. Vastly preferable is capturing the high-speed bits themselves, either through a sampling scope capable of storing a large number of samples (say 32 ksamples or more, something we lacked), or through a custom circuit like a 16:1 demultiplexer which would allow bits to be captured on a fast logic analyzer [Jen95]. Using an FFT to find the spectrum would almost certainly yield higher SNR than is found on a spectrum analyzer.

Overall, we were pleased that our first-cut deisgn functioned, but we would make a number of changes in a redesign:

- We would design appropriate impedance matching networks for both the input and clock.

- Redesigning the emitter followers in the preamp and half-latch in §2.2 can improve the preamp corner frequency to almost 20GHz from 4.6GHz and the half-latch regeneration time to $\tau_{rg} = 10.9$ps from 17.4ps. This gives fewer metastability problems and sharper transitions on the latch outputs, both of which can add one bit or more of performance.

- Some degree of performance loss caused by quantizer metastability could be alleviated by taking the output bit from the one-sample delayed latch output rather than the M/S output as we did in our design. This would provide the regeneration of two additional half-latches, which would lead to less of a tail on the right side of the zero-crossing histogram in Figure 7.33 and hence a somewhat lower white noise floor would be observed on a spectrum analyzer.

- We discovered after fabrication that we had scaled the DAC currents improperly. Figure 7.40a plots a histogram of the voltage at the input of the quantizer from a SPICE simulation; the maximum quantizer input magnitude is a mere 10 mV. Figure 7.40b shows the FFT of 4096 output bits, and it is readily apparent that the in-band noise is flat rather than shaped. We are clocking the quantizer (which is an ECL-style latched comparator) at 4 GHz in a process with $f_T = 50$ GHz, and this is aggressive enough that there needs to be a much larger voltage swing at the quantizer input for it to reliably switch. SPICE can be used to show that the solution to this problem lies in lowering the 2 kΩ emitter degeneration resistors in the DACs, which provides more DAC current and more voltage swing at the expense of higher power consumption.

- The previous two points could be rendered moot by including a method of capturing the digital bits.

- For manufacturability, we would likely replace the fixed resonator capacitors with varactors which could be tuned to counter process variations. To tune them, we could clock the modulator with an on-chip VCO also containing a varactor; tuning of both varactors could perhaps be accomplished simultaneously.

- Likewise, instead of scaling DAC currents with tunable voltages, we might consider replacing the bipolar sources with ratioed NMOS transistors to provide the appropriate feedback scaling for best noise-shaping. This obviates the need for DAC current tuning circuitry.

If all these changes worked to best effect, modulator DR would be limited by the G_{g1} transconductor to 52 dB or so. We feel that with careful redesign of this transconductor, we could achieve 60 dB (10 bits) of dynamic range. The

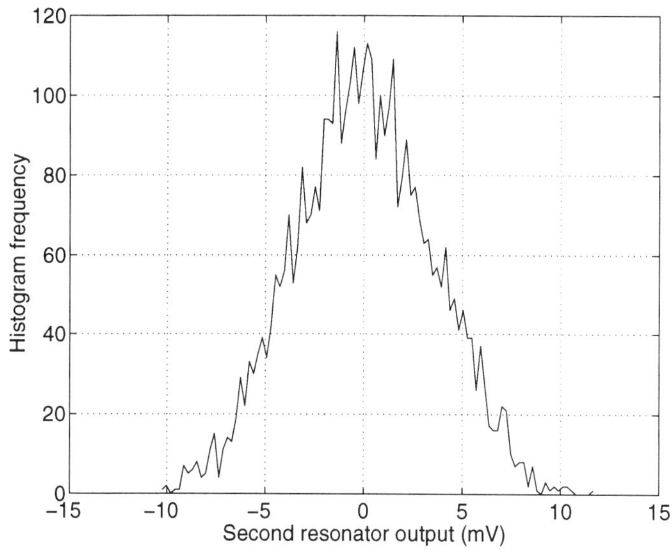

Figure 7.40a. Quantizer input voltage histogram.

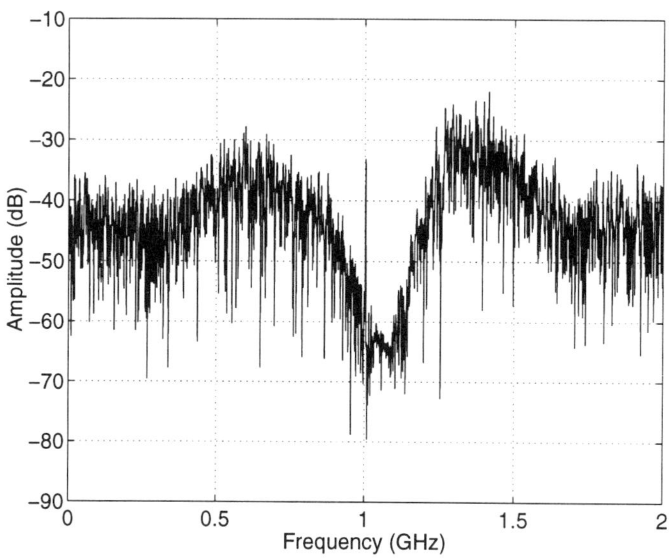

Figure 7.40b. SPICE output spectrum.

authors of [Jay97] concluded a similar performance level could be achieved in their 3.2 GHz $f_s/4$ modulator with a 25 MHz bandwidth.

5. SUMMARY

Table 7.3 summarizes the performance achieved by this high-speed BP

Table 7.3. Modulator performance summary.

Process	0.5μm SiGe HBT
Die area with pads	2.4mm × 1.6mm
Die area without pads	1.6mm × 0.85mm
Supply voltage	5V
Sampling frequency	4GHz
Signal bandwidth	20MHz
Oversampling ratio	100
Dynamic range	40dB
Peak SNR	37dB
Peak SFDR	48dB
Power consumption	450mW

$\Delta\Sigma$M. In a redesign, the output bit should be taken from the delayed latch output, the emitter followers should be optimized for speed, and the resonator output voltages should be scaled up; we estimate these changes would improve dynamic range from 6.3 bits to about 9 bits in a 20MHz bandwidth. It is our feeling that with further careful design of the input transconductor for low noise and higher linearity, a resolution of 10 bits might be achievable. A second-cut of our design would also include a method of capturing the output bits for off-line FFTs.

APPENDIX 7.A: Measurement Aids

In this appendix we plot graphs from SPICE that allow us to estimate important parameters from the fabricated $f_s/4$ BP modulator.

General transistor dc characteristics are plotted in Figures 7.A.1a and 7.A.1b. We show collector current against V_{BE} and V_{CE} for two commonly-used transistor sizes in this design, 5μm × 0.5μm and 20μm × 0.5μm.

Figures 7.A.2a and 7.A.2b show how varying the multi-tanh control voltages V_G and V_Q affects the transconductance G_g and G_q actually delivered.

Figure 7.A.3a contains the bias circuit for V_{EN} which sets the current in the latching stages, and Figure 7.A.3b plots the current in the emitter resistors of that circuit.

Figure 7.A.4 illustrates the current through one of the R_{DAC} resistors in Figure 7.21 as a function of the base voltage at the current-source transistor;

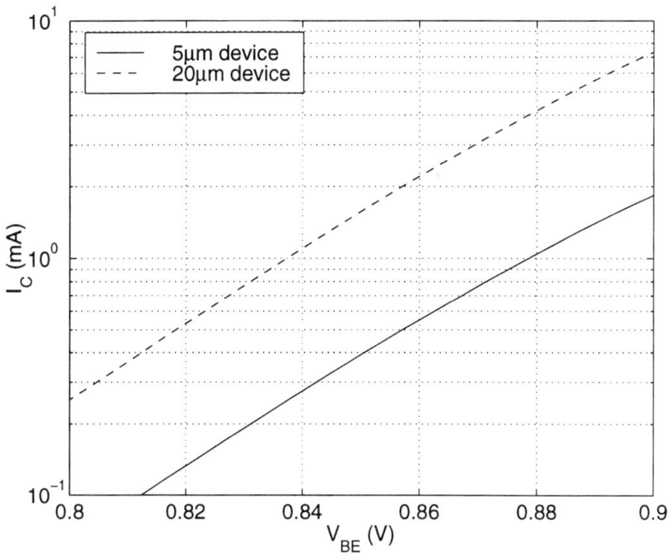

Figure 7.A.1a. BJT characteristics: I_C vs. V_{BE}.

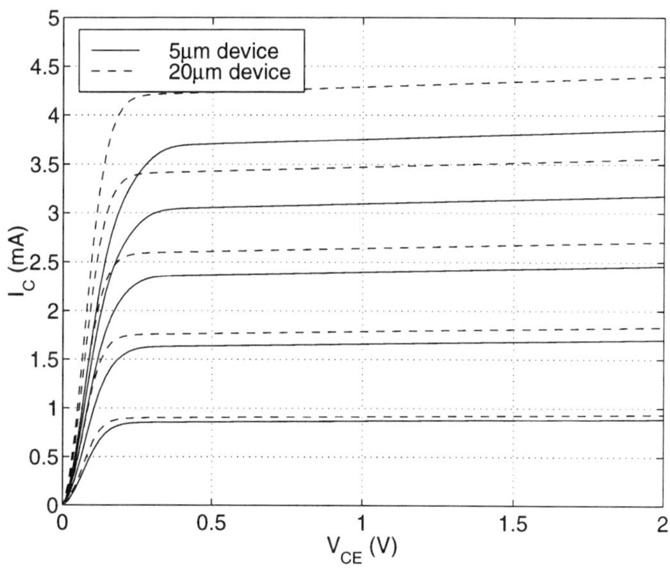

Figure 7.A.1b. BJT characteristics: I_C vs. V_{CE}.

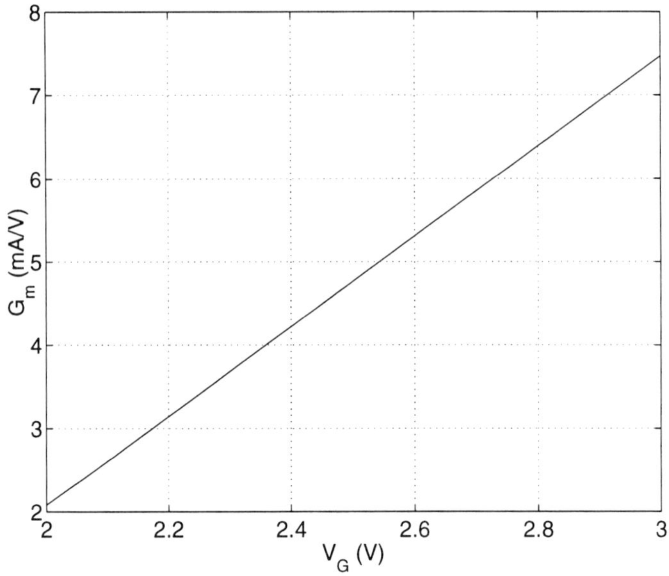

Figure 7.A.2a. G_g against V_G.

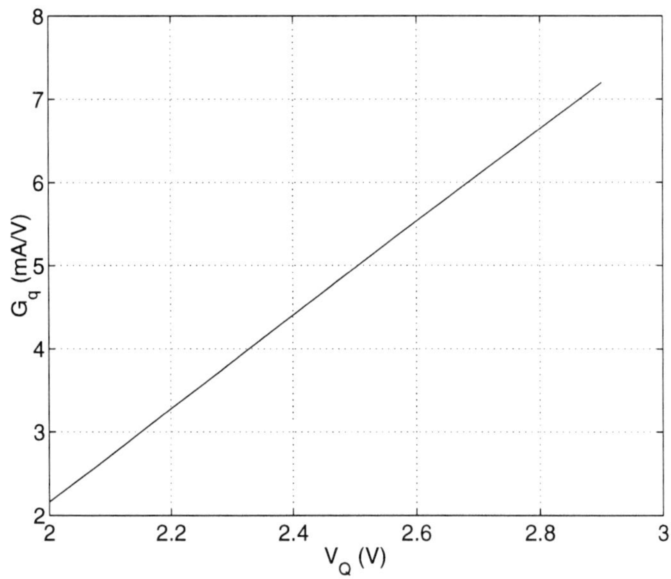

Figure 7.A.2b. G_q against V_Q.

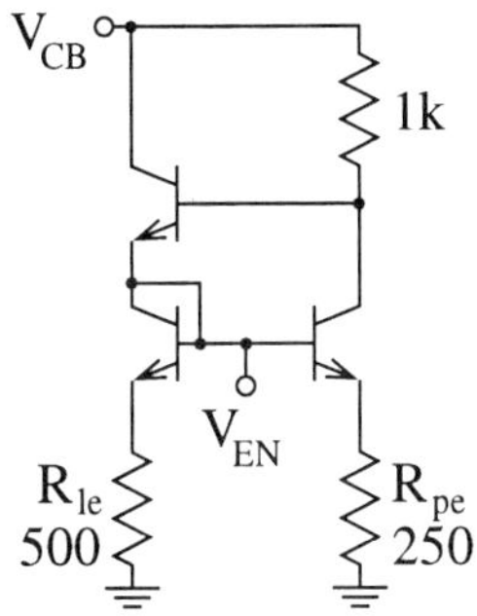

Figure 7.A.3a. V_{EN} bias circuit.

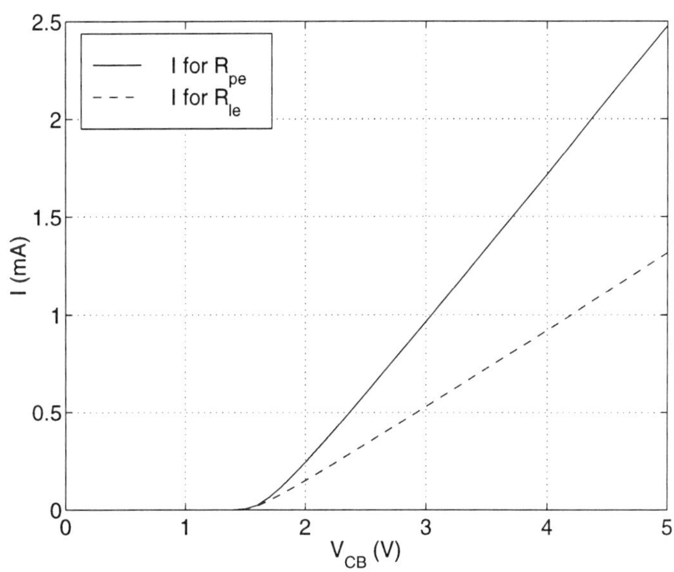

Figure 7.A.3b. Current through bias transistor emitter resistors.

the total amount of current switched is found by finding I_{k+} and I_{k-} for each of V_{k+} and V_{k-} separately from the graph, then subtracting the values.

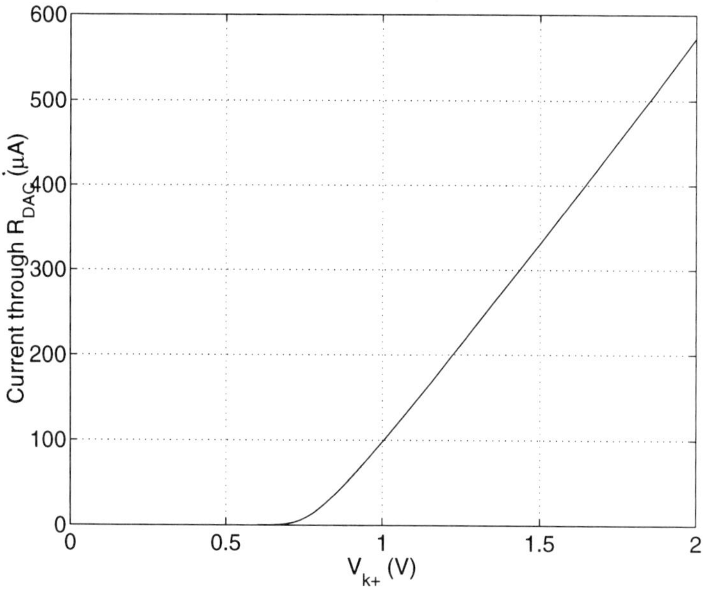

Figure 7.A.4. DAC current vs. control voltage.

Chapter 8

CONCLUSIONS

In this section, we summarize the work presented in Chapters 1–7 of this monograph, and we also present our conclusions regarding the usefulness of high-speed CT $\Delta\Sigma$M for ADC. Finally, we list some topics for future research.

1. SUMMARY OF RESULTS

In Chapter 1, we explained the basic operation of $\Delta\Sigma$Ms and listed the major design choices that exist. CT $\Delta\Sigma$Ms had several advantages over DT modulators: faster clocking, better virtual grounds, free antialiasing, and the possibility of using physical integrating devices as the first stage. Key for measuring performance of a $\Delta\Sigma$M was the power spectrum, from which SNR and DR can be found. To generate the power spectrum, time-domain simulation had to be employed; a tradeoff existed between simulation rapidity and realism.

We started Chapter 2 by explaining the ideal design of CT $\Delta\Sigma$Ms through the use of the impulse-invariant transform and a particular DAC pulse shape. We showed how to use tables or Matlab to design both low pass and band pass modulators, where for the latter we recommended using two different types of DAC so that purely-band pass CT resonators could be used as circuit blocks. We showed that CT $\Delta\Sigma$Ms have some built-in antialiasing because the sampling is done at the quantizer rather than at the input as in a DT $\Delta\Sigma$M, and we explored what happens when DAC pulse rise and fall times are unequal. Finally, we gave a SPICE-based procedure for choosing DAC feedback currents to implement a certain noise-shaping transfer function, and we explained the dual of this procedure where z-domain extraction is used to find the noise transfer function actually implemented in a CT $\Delta\Sigma$M.

Chapter 3 was a literature review of two major subjects: nonidealities in $\Delta\Sigma$M, and the important references in CT $\Delta\Sigma$M. We explained how common DT $\Delta\Sigma$M nonidealities are manifested in CT $\Delta\Sigma$Ms, and we demonstrated that published GHz-clocking CT $\Delta\Sigma$Ms typically achieve noise shaping only until an OSR of 15 or so.

The phenomenon of excess loop delay was the topic of Chapter 4. We used root locus to show how excess delay degrades loop stability, and we simulated the effect of excess delay on the DR of many different orders and types of low pass and band pass modulators. The out-of-band gain in high-order CT $\Delta\Sigma$Ms was a parameter which could be chosen to give a certain degree of performance for a given loop delay, though it had a larger effect in low pass modulators than band pass. Excess delay problems were circumvented by using RZ DAC pulses instead of NRZ, by tuning the DAC feedback levels, and by using additional feedbacks.

Chapter 5 studied two related CT $\Delta\Sigma$M problems: clock jitter and quantizer metastability. Both had the effect of varying the width of the feedback DAC pulses, which modulated noise from outside the signal band into the signal band. We concluded that a fully-integrated modulator with a typical GHz-speed VCO was unlikely to be performance limited by that VCO's phase noise, and that using another half latch in the feedback path for additional regeneration was essential to avoiding metastability problems at high speeds. Even still, clocking a modulator faster than 5% of the transistor f_T was not recommended.

A procedure for choosing the main parameters in a fourth-order $f_s/4$ band pass modulator was outlined in Chapter 6. We noted that designing the first transconductor for high DR at high speeds is a major challenge, but one that must be met in order not to limit the DR of the overall modulator too severely. Apart from the input transconductor, the other major design choices we explored were the tank components and positive feedback transconductor, DAC current levels for appropriate signal scaling, and transconductor linearity.

Chapter 7 was about a 4GHz-clocking band pass $\Delta\Sigma$M for A/D conversion of 1GHz radio signals. Detailed circuit block descriptions and simulation results were given, and we presented test results on many aspects of modulator performance. Clock jitter turned out to be the major reason for poor performance; though we proposed several possible sources of it, we were unable to pinpoint its most likely cause. A spectrum analyzer was not the way to measure high-speed modulator performance: capturing the bits with a deep sampling oscilloscope or through a demux and fast logic analyzer would have been preferable. With a redesign, we estimated we could achieve 10-bit performance in a 20MHz bandwidth rather than the 6.3 bits we observed.

2. USEFULNESS OF HIGH-SPEED CT $\Delta\Sigma$M

We close by addressing the following question: is wideband UHF ADC practical with high-speed $\Delta\Sigma$Ms? For bandwidths into the tens of MHz, it is a major challenge to get resolutions much above 10 bits with these circuits. The use of BP modulators followed by digital mixing and filtering is a tempting architecture choice, but a converter that works on an entire cellular band while maintaining a high-enough DR seems like a virtual impossibility given our test results and those of others.

Apart from jitter problems, thermal noise is a serious problem at high conversion bandwidths. Consider an audio converter with a bandwidth of 44 kHz that achieves 16-bit resolution with a $\Delta\Sigma$M whose performance is limited by white thermal noise rather than shaped quantization noise. A similar thermal noise power level in a modulator clocking 1000 times faster with the same OSR (i.e., with a 44 MHz bandwidth) would have 30 dB more in-band thermal noise and a resolution of only 11 bits. Perhaps one solution is to use an AGC amplifier in front of the $\Delta\Sigma$M; another is to perform the conversion at a much lower IF where circuit nonidealities and jitter problems can be more easily overcome.

This is not at all to say high-speed $\Delta\Sigma$M is impractical in all cases. Three published circuits of interest are as follows:

1. Narrowband high-speed $\Delta\Sigma$M: [Rag97] reports a 15-bit ADC with a 370 kHz bandwidth programmable from 0 to 70 MHz. Even with white in-band noise, oversampling still improves resolution, albeit at the slow rate of 0.5 bits per octave. With enough oversampling, high resolution can be achieved.

2. Hybrid mixer/modulator: [Mor98] uses two low pass modulators for I and Q channel demodulation with wideband mixers built right into the front end. A desired narrowband channel can simultaneously be mixed to dc and converted to digital for filtering. Once again, high-speed $\Delta\Sigma$Ms are good for high-resolution narrowband conversion, and the modulator has no complicated tuning circuitry: channel selection is done by tuning a VCO.

3. Wideband ADC with a hybrid architecture: [Bro97] reports a 1.25 MHz modulator with 16-bit resolution that uses a multibit $\Delta\Sigma$M front end oversampled eight times, and a Nyquist-rate pipelined back end. The strengths of the individual circuits are exploited, the $\Delta\Sigma$M for high DR but limited bandwidth and the pipelined converter for high bandwidth but limited DR.

$\Delta\Sigma$Ms are complicated circuits, particularly at high sampling rates. In our opinion, it is more useful to place the emphasis on finding architectures which exploit the strengths of $\Delta\Sigma$M rather than pushing the clock speed of the modulator. As clock speeds increase significantly (20GHz and beyond, perhaps), surely other nonidealities will start to degrade performance: substrate noise coupling, transmission line effects, etc. Designing even simple circuits like multiplexers at these speeds poses a number challenges; CT $\Delta\Sigma$Ms are complicated circuits, which amplifies the argument favoring cleverness rather than clock speed.

3. FUTURE WORK

There are, of course, a number of areas in which the state-of-the-art for CT ΔΣMs could well be advanced, thus improving their usefulness in a wider range of applications. The following problems remain to be studied and solved; they are listed in order beginning with what we consider to be most important.

Multibit DAC A working high-speed multibit design could be a significant breakthrough: as we said in Chapter 4 §5., not only are multibit modulators higher resolution and more stable, but they improve clock jitter sensitivity too. Can fast DEM be made to work?

Calibration and tuning How does one tune a high-speed CT ΔΣM for maximum DR over process and temperature variations, either dynamically or off-line? For production parts, this is essential. Yet high-speed circuits are best when kept simple, and tuning will add complexity. This seems a tough problem to tackle.

System identification We attempted to give a method for rapid identification of nonidealities in a ΔΣM in Chapter 2 §4.2. Can this be improved? That is, can we come up with a way to pinpoint modulator problem areas *rapidly* and *accurately*? Moreover, can we find a way to apply it to a real modulator in the lab, rather than just in simulation?

Power consumption Can power be reduced through non-bipolar circuits or lower supply voltages while maintaining speed?

Higher modulator order Is it worth going to a higher-order design for high-speed ΔΣM for the resolution gained? High-order audio converters often include reset circuitry that activates when modulator overload is sensed [dS90]; can such circuits be included in a GHz-speed design? Are they necessary?

There is still plenty of exciting work left to do in the field of high-speed CT ΔΣM.

References

[Ada86] R. Adams. Design and implementation of an audio 18-bit analog-to-digital converter using oversampling techniques. *J. Audio Eng. Society*, pages 153–166, March/April 1986.

[Ada98] R. Adams, K. Nguyen, and K. Sweetland. A 113dB SNR oversampling DAC with segmented noise-shaped scrambling. In *Int. Solid-State Circ. Conf. Technical Digest*, pages 62–63, 1998.

[Asi91] R. L. Asibal and P. E. Allen. Design issues in GaAs oversampled A/D converters. In *Proc. Int. Symp. Circ. Syst.*, volume 3, pages 1841–1844, 1991.

[Azi96] P. M. Aziz, H. V. Sorensen, and J. Van Der Spiegel. An overview of sigma–delta converters. *IEEE Signal Proc. Mag.*, pages 61–84, January 1996.

[Bai95] R. T. Baird and T. S. Fiez. Linearity enhancement of multibit $\Delta\Sigma$ A/D and D/A converters using data weighted averaging. *IEEE Trans. Circ. Syst. II*, pages 753–762, December 1995.

[Bai96] R. T. Baird and T. S. Fiez. A low oversampling ratio 14-b 500-kHz $\Delta\Sigma$ ADC with a self-calibrated multibit DAC. *IEEE J. Solid-State Circ.*, pages 312–320, March 1996.

[Baz96] S. Bazarjani. *Mixed Analog-Digital Design Considerations in Deep Submicron CMOS Technologies*. PhD thesis, Carleton University, 1996.

[Ben48] W. R. Bennett. Spectra of quantized signals. *Bell Syst. Tech. J.*, pages 446–472, July 1948.

[Ben97] P. Benabes, M. Keramat, and R. Kielbasa. A methodology for designing continuous-time sigma-delta modulators. In *IEEE European Design and Test Conf.*, pages 46–50, 1997.

[Ber92] A. Berkovitz and I. Rusnak. FFT processing of randomly sampled harmonic signals. *IEEE Trans. Signal Proc.*, pages 2816–2819, November 1992.

[Ber96] B. Bereza. *Dynamic Range and Bandwidth Limitations in Sampled-Data and Clocked Continuous-Time BiCMOS Sigma-Delta Modulators*. PhD thesis, Carleton University, 1996.

[Bos88] B. E. Boser and B. A. Wooley. The design of sigma–delta modulation analog-to-digital converters. *IEEE J. Solid-State Circ.*, pages 1298–1308, December 1988.

[Bro90] J. E. C. Brown, M. Alexander, and D. F. Bowers. Mixed-mode simluation of a continuous-time $\Sigma\Delta$ ADC. In *Proc. Int. Symp. Circ. Syst.*, volume 3, pages 1915–1918, 1990.

[Bro97] T. L. Brooks, D. H. Robertson, D. F. Kelly, A. Del Muro, and S. W. Harston. A 16b $\Sigma\Delta$ pipeline ADC with 2.5MHz output data-rate. In *Int. Solid-State Circ. Conf. Technical Digest*, pages 208–209, 1997.

[Can85] J. C. Candy. A use of double integration in sigma delta modulation. *IEEE Trans. Communications*, pages 249–258, March 1985.

[Can92a] J. C. Candy and G. C. Temes, editors. *Oversampling Delta-Sigma Data Converters: Theory, Design, and Simulation*. IEEE Press, New York, 1992.

[Can92b] J. C. Candy and G. C. Temes. Oversampling methods for A/D and D/A conversion. In J. C. Candy and G. C. Temes, editors, *Oversampling Delta-Sigma Data Converters: Theory, Design, and Simulation*, pages 1–28. IEEE Press, New York, 1992.

[Can97] J. C. Candy. An overview of basic concepts. In S. R. Norsworthy, R. Schreier, and G. C. Temes, editors, *Delta-Sigma Data Converters: Theory, Design, and Simulation*, chapter 1. IEEE Press, New York, 1997.

[Car87] L. R. Carley. An oversampling analog-to-digital converter topology for high-resolution signal acquisition systems. *IEEE Trans. Circ. Syst.*, pages 83–90, January 1987.

[Cha90] K. C.-H. Chao, S. Nadeem, W. L. Lee, and C. G. Sodini. A higher order topology for interpolative modulators for oversampled A/D converters. *IEEE Trans. Circ. Syst.*, pages 309–318, March 1990.

[Cha92] K. T. Chan and K. W. Martin. Components for a GaAs delta-sigma modulator oversampled analog-to-digital converter. In *Proc. Int. Symp. Circ. Syst.*, pages 1300–1303, 1992.

[Che94] J. A. Cherry. Distortion analysis of weakly nonlinear filters using Volterra series. Master's thesis, Carleton University, 1994.

[Che97] J. A. Cherry, W. M. Snelgrove, and P. Schvan. Signal-dependent timing jitter in continuous-time $\Sigma\Delta$ modulators. *Electronics Letters*, 33(13):1118–1119, June 1997.

[Che98a] J. A. Cherry and W. M. Snelgrove. Approaches to simulating continuous-time $\Sigma\Delta$ modulators. In *Proc. Int. Symp. Circ. Syst.*, volume 1, pages 587–590, 1998.

[Che98b] J. A. Cherry and W. M. Snelgrove. Loop delay and jitter in continuous-time $\Sigma\Delta$ modulators. In *Proc. Int. Symp. Circ. Syst.*, volume 1, pages 596–599, 1998.

[Che98c] J. A. Cherry and W. M. Snelgrove. On the characterization and reduction of distortion in bandpass filters. *IEEE Trans. Circ. Syst. I*, pages 523–537, May 1998.

[Che99a] J. A. Cherry and W. M. Snelgrove. Clock jitter and quantizer metastability in continuous-time delta–sigma modulators. *IEEE Trans. Circ. Syst. II*, pages 661–676, June 1999.

[Che99b] J. A. Cherry and W. M. Snelgrove. Excess loop delay in continuous-time delta–sigma modulators. *IEEE Trans. Circ. Syst. II*, pages 376–389, April 1999.

[Com91] V. Comino, M. S. J. Steyaert, and G. C. Temes. A first-order current-steering sigma-delta modulator. *IEEE J. Solid-State Circ.*, pages 176–183, March 1991.

[Cur95] K. W. Current, J. F. Parker, and W. J. Hardraker. On behavioral modeling of analog and mixed-signal circuits. In *Asilomar Conf. on Signals, Systems, and Computers*, volume 1, pages 264–268, 1995.

[Cut60] C. C. Cutler. Transmission systems employing quantization. U.S. Patent no. 2,927,962, 1960.

[Dau98] P. L. Dauphinée. Private communication, 1998.

[Dia92a] V. F. Dias, G. Palmisano, and F. Maloberti. Noise in mixed continuous-time switched-capacitor sigma-delta modulators. *Proc. IEE Part G*, pages 680–684, December 1992.

[Dia92b] V. F. Dias, G. Palmisano, P. O'Leary, and F. Maloberti. Fundamental limitations of switched-capacitor sigma-delta modulators. *Proc. IEE Part G*, pages 27–32, February 1992.

[Dia94] V. F. Dias, G. Palmisano, and F. Maloberti. Harmonic distortion in SC sigma-delta modulator. *IEEE Trans. Circ. Syst. I*, pages 326–329, April 1994.

[Don97] Y. Dong and A. Opal. Efficient monte-carlo thermal noise simulation for $\Sigma\Delta$ modulators. In *Proc. Custom Integrated Circ. Conf.*, pages 499–502, 1997.

[Don98a] Y. Dong and A. Opal. Fast time-domain noise simulation of sigma-delta converters and periodically switched linear networks. In *Proc. Int. Symp. Circ. Syst.*, volume 6, pages 114–118, 1998.

[Don98b] Y. Dong and A. Opal. Dithering effect simulation for sigma-delta modulators with the presence of thermal noise. In *CITO Inaugural Researcher Retreat Workbook*, May 12-14, 1998.

[dS90] B. P. del Signore, D. A. Kerth, N. S. Sooch, and E. J. Swanson. A monolithic 20-b delta-sigma A/D converter. *IEEE J. Solid-State Circ.*, pages 1311–1317, December 1990.

[Erb96] M. Erbar, M. Rieger, and H. Schemmann. A 1.28-GHz sigma-delta modulator for video A/D conversion. In *Int. Conf. Consumer Elec.*, pages 78–79, 1996.

[Fee91] O. Feely and L. O. Chua. The effect of integrator leak in Σ–Δ modulation. *IEEE Trans. Circ. Syst.*, pages 1293–1305, November 1991.

[Fen94] S. Feng, J. Sauerer, and D. Seitzer. Implementation of GaAs E/D HEMT analog components for oversampling analog/digital conversion. In *GaAs Int. Circ. Symp.*, pages 228–231, 1994.

[Gai89] P. H. Gailus, W. J. Turney, and F. R. Yester Jr. Method and arrangement for a sigma delta converter for bandpass signals. U.S. Patent no. 4,857,928, issued Aug. 15, 1989.

[Gal96] I. Galton. Noise-shaping D/A conversion for $\Delta\Sigma$ modulation. In *Proc. Int. Symp. Circ. Syst.*, volume 1, pages 441–444, 1996.

[Gao97a] W. Gao, O. Shoaei, and W. M. Snelgrove. Excess loop delay effects in continuous-time delta-sigma modulators and the compensation solution. In *Proc. Int. Symp. Circ. Syst.*, volume 1, pages 65–68, 1997.

[Gao97b] W. Gao and W. M. Snelgrove. A 950MHz second-order integrated LC bandpass $\Delta\Sigma$ modulator. In *Proc. VLSI Circ. Symp.*, pages 111–112, 1997.

[Gao98a] W. Gao, J. A. Cherry, and W. M. Snelgrove. A 4GHz fourth-order SiGe HBT band pass $\Delta\Sigma$ modulator. In *Proc. VLSI Circ. Symp.*, pages 174–175, 1998.

[Gao98b] W. Gao and W. M. Snelgrove. A 950-MHz IF second-order integrated LC bandpass delta–sigma modulator. *IEEE J. Solid-State Circ.*, pages 723–732, May 1998.

[Gar86] F. M. Gardner. A transformation for digital simulation of analog filters. *IEEE Trans. Communications*, pages 676–680, July 1986.

[Gil98] B. Gilbert. The multi-tanh principle: A tutorial overview. *IEEE J. Solid-State Circ.*, pages 2–17, January 1998.

[Gos88] A. Gosslau and A. Gottwald. Optimization of a sigma-delta modulator by the use of a slow ADC. In *Proc. Int. Symp. Circ. Syst.*, pages 2317–2320, 1988.

[Gos90] A. Gosslau and A. Gottwald. Linearization of a sigma-delta modulator by a proper loop delay. In *Proc. Int. Symp. Circ. Syst.*, volume 1, pages 364–367, 1990.

[Gra90] R. M. Gray. Quantization noise spectra. *IEEE Trans. Information Theory*, pages 1220–1244, November 1990.

[Gra93] P. R. Gray and R. G. Meyer. *Analysis and Design of Analog Integrated Circuits.* John Wiley & Sons, third edition, 1993.

[Gre86] R. Gregorian and G. C. Temes. *Analog MOS Integrated Circuits for Signal Processing.* John Wiley & Sons, New York, 1986.

[Haj98] A. Hajimiri and T. H. Lee. A general theory of phase noise in electrical oscillators. *IEEE J. Solid-State Circ.*, pages 179–194, February 1998.

[Hal92] B. Hallgren. Design of a second order CMOS sigma-delta A/D converter with a 150 MHz clock rate. In *Proc. European Solid-State Circ. Conf.*, pages 103–106, 1992.

[Han98] D. C. Hanselman and B. Littlefield. *Mastering MATLAB 5: A Comprehensive Tutorial and Reference.* Prentice Hall, 1998.

[Har78] F. J. Harris. On the use of windows for harmonic analysis with the discrete Fourier transform. *Proc. IEEE*, pages 51–83, January 1978.

[Har90] S. Harris. The effects of sampling clock jitter on Nyquist sampling analog-to-digital converters, and on oversampling delta–sigma ADCs. *J. Audio Eng. Society*, pages 537–542, July/August 1990.

[Har95] D. L. Harame, J. H. Comfort, J. D. Cressler, E. F. Crabbe, J. Y.-C. Sun, B. S. Meyerson, and T. Rice. Si/SiGe epitaxial-base transistors—Part I: Materials, physics, and circuits. *IEEE Trans. Electron Devices*, pages 455–468, March 1995.

[Hau86] M. W. Hauser and R. W. Brodersen. Circuit and technology considerations for MOS delta-sigma A/D converters. In *Proc. Int. Symp. Circ. Syst.*, pages 1310–1315, 1986.

[Hau90] M. W. Hauser. Technology scaling and performance limitations in delta-sigma analog-digital converters. In *Proc. Int. Symp. Circ. Syst.*, pages 356–359, 1990.

[Hau91] M. W. Hauser. Principles of oversampling A/D conversion. *J. Audio Eng. Society*, pages 3–26, January/February 1991.

[Hay86] T. Hayashi, Y. Inabe, K. Uchimura, and T. Kimura. A multi stage delta–sigma modulator without double integration loop. In *Int. Solid-State Circ. Conf. Technical Digest*, pages 182–183, 1986.

[Hor89] J. U. Horstmann, H. W. Eichel, and R. L. Coates. Metastability behavior of CMOS ASIC flip-flops in theory and test. *IEEE J. Solid-State Circ.*, pages 146–157, February 1989.

[Hor90] U. Horbach. Design of a 20 bit sigma-delta A/D-converter for audio applications. In *Proc. Int. Symp. Circ. Syst.*, volume 4, pages 2789–2792, 1990.

[Ino62] H. Inose, Y. Yasuda, and J. Murakami. A telemetering system by code modulation — Δ–Σ modulation. *IRE Trans. Space Elec. Telemetry*, pages 204–209, September 1962.

[Ino63] H. Inose and Y. Yasuda. A unity bit coding method by negative feedback. *Proc. IEEE*, pages 1524–1535, November 1963.

[Jay97] A. Jayaraman, P. Asbeck, K. Nary, S. Beccue, and K.-C. Wang. Bandpass delta-sigma modulator with 800 MHz center frequency. In *GaAs Int. Circ. Symp.*, pages 95–98, 1997.

[Jen94] J. F. Jensen, A. E. Cosand, W. E. Stanchina, R. H. Walden, T. Lui, Y. K. Brown, M. Montes, K. Elliot, and C. G. Kirkpatrick. Double heterostructure InP HBT technology for high resolution A/D converters. In *GaAs Int. Circ. Symp.*, pages 224–227, 1994.

[Jen95] J. F. Jensen, G. Raghavan, A. E. Cosand, and R. H. Walden. A 3.2-GHz second-order delta-sigma modulator implemented in InP HBT technology. *IEEE J. Solid-State Circ.*, pages 1119–1127, October 1995.

[Jen98] H. T. Jensen and I. Galton. A reduced-complexity mismatch-shaping DAC for delta-sigma data converters. In *Proc. Int. Symp. Circ. Syst.*, volume 1, pages 504–507, 1998.

[Kaw96] S. Kawahito, A. Yamasawa, S. Koga, Y. Takodoro, K. Mizumo, and O. Tabata. Digital interface circuits using sigma–delta modulation for integrated micro fluxgate magnetic sensors. In *Proc. Int. Symp. Circ. Syst.*, volume 4, pages 340–343, 1996.

[Kes90a] W. Kester. Part 1: Flash ADCs provide the basis for high-speed conversion. *E. D. N.*, 35(1):101–109, January 1990.

[Kes90b] W. Kester. Part 3: Measure flash-ADC performance for trouble-free operation. *E. D. N.*, 35(3):103–114, February 1990.

[Koc86] R. Koch and B. Heise. A 120kHz sigma/delta A/D converter. In *Int. Solid-State Circ. Conf. Technical Digest*, pages 138–139, 1986.

[Kwa96] T. Kwan, R. Adams, and R. Libert. A stereo multibit $\Sigma\Delta$ DAC with asynchronous master clock interface. *IEEE J. Solid-State Circ.*, pages 1881–1887, December 1996.

[Lar88] L. E. Larson, T. Cataltepe, and G. C. Temes. Multibit oversampled Σ–Δ A/D convertor with digital error correction. *Electronics Letters*, pages 1051–1052, August 1988.

[Lee66] D. B. Leeson. A simple model of feedback oscillator noise spectra. *Proc. IEEE*, pages 329–330, February 1966.

[Lee92] I. Lee, J. Yang, and W. Kim. An analysis of clock feedthrough noise in bipolar comparators. In *Proc. Int. Symp. Circ. Syst.*, volume 3, pages 1392–1395, 1992.

[Lee98] T. H. Lee. *The Design of CMOS Radio-Frequency Integrated Circuits.* Cambridge University Press, 1998.

[Lib93] V. Liberali, V. F. Dias, M. Ciapponi, and F. Maloberti. TOSCA: A simulator for switched-capacitor noise-shaping A/D converters. *IEEE Trans. Computer-Aided Design*, pages 1376–1386, September 1993.

[Lu95] C. Lu, M. Lemkin, and B. E. Boser. A monolithic surface micromachined accelerometer with digital output. *IEEE J. Solid-State Circ.*, pages 1367–1373, December 1995.

[Mat87] Y. Matsuya, K. Uchimura, A. Iwata, T. Kobayashi, M. Ishikawa, and T. Yoshitome. A 16-bit oversampling A-to-D conversion technology using triple-integration noise shaping. *IEEE J. Solid-State Circ.*, pages 921–929, December 1987.

[Med94] F. Medeiro, B. Pérez-Verdú, A. Rodríguez-Vázquez, and J. L. Huertas. Modeling opamp-induced harmonic distortion for switched-capacitor $\Sigma\Delta$ modulator design. In *Proc. Int. Symp. Circ. Syst.*, volume 5, pages 445–448, 1994.

[Med95] F. Medeiro, B. Pérez-Verdú, A. Rodríguez-Vázquez, and J. L. Huertas. A vertically integrated tool for automated design of $\Sigma\Delta$ modulators. *IEEE J. Solid-State Circ.*, pages 762–772, July 1995.

[Mit95] R. Mittal and D. J. Allstot. Low-power high-speed continuous-time Σ–Δ modulators. In *Proc. Int. Symp. Circ. Syst.*, volume 1, pages 183–186, 1995.

[Miy97] T. Miyashita, A. Olmos, M. Nihei, and Y. Watanabe. 5 GHz $\Sigma\Delta$ analog-to-digital converter with polarity alternating feedback comparator. In *GaAs Int. Circ. Symp.*, pages 91–93, 1997.

[Mor98] M. A. Morin. A 1.6 Gb/s delta-sigma modulator with integrated wideband mixer for RF applications. In *Proc. Bipolar/BiCMOS Circ. Tech. Meeting*, pages 148–151, 1998.

[Nar94] K. R. Nary, S. Beccue, R. Nubling, R. Pierson, K.-C. Wang, P. Zampardi, and A. Jayaraman. Second order $\Delta\Sigma$ modulators using AlGaAs/GaAs HBTs. In *GaAs Int. Circ. Symp.*, pages 232–235, 1994.

[Ned95] J. Nedved, J. Vanneuville, D. Gevaert, and J. Sevenhans. A transistor-only switched current sigma-delta A/D converter for a CMOS speech codec. *IEEE J. Solid-State Circ.*, pages 819–822, July 1995.

[Nor97] S. R. Norsworthy, R. Schreier, and G. C. Temes, editors. *Delta-Sigma Data Converters: Theory, Design, and Simulation*. IEEE Press, New York, 1997.

[Nys96] O. Nys and R. Henderson. A monolithic 19-bit 800Hz low power multi-bit sigma-delta CMOS ADC using data weighted averaging. In *Proc. European Solid-State Circ. Conf.*, pages 252–255, 1996.

[Oli98] O. Oliaei and H. Aboushady. Jitter effects in continuous-time $\Sigma\Delta$ modulation with delayed return-to-zero feedback. In *Proc. IEEE Int. Conf. on Electronics Circ. Syst.*, volume 1, pages 351–354, 1998.

[Olm98] A. Olmos, T. Miyashita, M. Nihei, E. Charry, and Y. Watanabe. A 5GHz continuous time sigma-delta modulator implemented in 0.4μm InGaP/InGaAs HEMT technology. In *Proc. Int. Symp. Circ. Syst.*, volume 1, pages 575–578, 1998.

[Opa96] A. Opal. Sampled data simulation of linear and nonlinear circuits. *IEEE Trans. Computer-Aided Design*, pages 295–307, March 1996.

[Pea87] T. H. Pearce and A. C. Baker. Analogue to digital conversion requirements for HF radio receivers. In *Proceedings of the IEE Colloquium on System Aspects and Applications of ADCs for Radar, Sonar, and Communications*, London, November 1987.

[Pre92] W. H. Press, S. A. Teukolsky, W. T. Vetterling, and B. P. Flannery. *Numerical Recipes in C.* Cambridge University Press, Cambridge, second edition, 1992.

[Pto97] U. C. Berkeley EECS Department. *The Almagest: Ptolemy 0.7 User's Manual*, 1997. `http://ptolemy.eecs.berkeley-.edu/papers/almagest`.

[Rag97] G. Raghavan, J. F. Jensen, R. H. Walden, and W. P. Posey. A bandpass $\Sigma\Delta$ modulator with 92dB SNR and center frequency continuously programmable from 0 to 70MHz. In *Int. Solid-State Circ. Conf. Technical Digest*, pages 214–215, 1997.

[Reb89] M. Rebeschini, N. van Bavel, P. Rakers, R. Greene, J. Caldwell, and J. Haug. A high-resolution CMOS sigma-delta A/D converter with 320 kHz output rate. In *Proc. Int. Symp. Circ. Syst.*, pages 246–249, 1989.

[Red94] D. Redfern. *The Maple Handbook.* Springer-Verlag, New York, 1994.

[Ris94] L. Risbo. *$\Sigma\Delta$ Modulators – Stability and Design Optimization.* PhD thesis, Technical University of Denmark, 1994.

[Sch89] R. Schreier and W. M. Snelgrove. Bandpass sigma-delta modulation. *Electronics Letters*, 25(23):1560–1561, November 1989.

[Sch91] R. Schreier. *Noise-Shaped Coding.* PhD thesis, Univeristy of Toronto, 1991.

[Sch93] R. Schreier. An empirical study of high-order single-bit delta-sigma modulators. *IEEE Trans. Circ. Syst. II*, pages 461–466, August 1993.

[Sch96a] R. Schreier and B. Zhang. Delta-sigma modulators employing continuous-time circuitry. *IEEE Trans. Circ. Syst. I*, pages 324–332, April 1996.

[Sch96b] P. Schvan. Unpublished work, 1996.

[Sho94] O. Shoaei and W. M. Snelgrove. Optimal (bandpass) continuous-time $\Sigma\Delta$ modulator. In *Proc. Int. Symp. Circ. Syst.*, volume 5, pages 489–492, 1994.

[Sho95] O. Shoaei and W. M. Snelgrove. A multi-feedback design for LC bandpass delta-sigma modulators. In *Proc. Int. Symp. Circ. Syst.*, volume 1, pages 171–174, 1995.

[Sho96] O. Shoaei. *Continuous-Time Delta-Sigma A/D Converters for High Speed Applications.* PhD thesis, Carleton Univeristy, 1996.

[Sho97] O. Shoaei and W. M. Snelgrove. Design and implementation of a tunable 40 MHz–70 MHz Gm-C bandpass $\Delta\Sigma$ modulator. *IEEE Trans. Circ. Syst. II*, pages 521–530, July 1997.

[Shu98] T. Shui, R. Schreier, and F. Hudson. Mismatch-shaping DAC for lowpass and bandpass multi-bit delta-sigma modulators. In *Proc. Int. Symp. Circ. Syst.*, volume 1, pages 352–355, 1998.

[Sim96] The Mathworks, Inc. *Using SIMULINK Version 2*, 1996.

[Sin95] F. W. Singor and W. M. Snelgrove. Switched-capacitor delta-sigma A/D modulation at 10.7MHz. *IEEE J. Solid-State Circ.*, pages 184–192, March 1995.

[Sne96] W. M. Snelgrove. *VLSI Circuits.* Unpublished, 1996.

[SPW92] Comdisco Systems, Inc. *SPW—The DSP Framework: User's Guide and Tutorial*, 1992.

[Thu91] A. M. Thurston, T. H. Pearce, and M. J. Hawksford. Bandpass implementation of the sigma delta A-D conversion technique. In *Int. Conf. on A.-D. and D.-A. Conversion*, pages 81–86, 1991.

[Tro93] G. Troster, H.-J. Dressler, H.-J. Golberg, W. Schardein, E. Zocher, A. Wedel, K. Schoppe, and J. Arndt. An interpolative bandpass converter on a 1.2μm BiCMOS analog/digital array. *IEEE J. Solid-State Circ.*, pages 471–477, April 1993.

[Ush94] G. Ushaw and S. McLaughlin. On the stability and configuration of sigma-delta modulators. In *Proc. Int. Symp. Circ. Syst.*, volume 5, pages 349–352, 1994.

[vdP94] R. van de Plassche. *Integrated Analog-to-Digital and Digital-to-Analog Converters.* Kluwer Academic Publishers, Dordrecht, NL, 1994.

[vdZ96] E. J. van der Zwan and E. C. Dijkmans. A 0.2-mW CMOS $\Sigma\Delta$ modulator for speech coding with 80dB dynamic range. *IEEE J. Solid-State Circ.*, pages 1873–1880, December 1996.

[vdZ97] E. J. van der Zwan. A 2.3mW CMOS $\Sigma\Delta$ modulator for audio applications. In *Int. Solid-State Circ. Conf. Technical Digest*, pages 220–221, 1997.

[Vee80] H. J. M. Veendrick. The behavior of flip-flops used as synchronizers and prediction of their failure rate. *IEEE J. Solid-State Circ.*, pages 169–176, April 1980.

[Wal96] L. Wall, T. Christiansen, and R. L. Schwartz. *Programming Perl.* O'Reilly & Associates, Sebastopol, CA, second edition, 1996.

[Wan92] H. Wang. A geometric view of $\Sigma\Delta$ modulations. *IEEE Trans. Circ. Syst. II*, pages 402–405, June 1992.

[Wil92] L. A. Williams, III and B. A. Wooley. MIDAS – a functional simulator for mixed digital and analog sampled data systems. In *Proc. Int. Symp. Circ. Syst.*, volume 5, pages 2148–2151, 1992.

[Yuk87] A. Yukawa. Constraints analysis for oversampling A-to-D converter structures on VLSI implementation. In *Proc. Int. Symp. Circ. Syst.*, pages 467–472, 1987.

Index